Facing the Climate Future Together

Voices for Advocacy and Action

Ingrid Rogers

Facing the Climate Future Together

Voices for Advocacy and Action

Ingrid Rogers

Cover photos by Matt Schlegel
Photos inside the book by Casey Cameron,
Ingrid Rogers, and Matt Schlegel
Artwork by Shirley Abraham, Karen Klimpert,
and Elaine Salinger
Poems and songs by Ingrid Rogers

For requests, contact IngridNRogers@gmail.com.

ISBN 978-1-938090-25-7

For Those
Who Have Spoken Up
In Defense of Our Planet

If we see ourselves as participants
in a cosmic garden of life
evolved over billions of years,
our regard for the universe will shift
from indifference, fear, and cynicism
to curiosity, love, and awe.

Contents

About This Book

You are probably concerned about the climate already. Something about this book caught your attention. Perhaps you are just contemplating what you might do in response to climate change, or maybe you have worked for years to protect the planet in some way and need encouragement to continue, or you are interested in learning what others are up to. Regardless, welcome to a community of concerned people.

As we ponder how to deal with the climate crisis, one of the most important realizations is that we are not alone. The challenges are huge, and it will require all of us to deal with them together. Every step toward healing the planet matters. Some actions have great impact, some reach large audiences, and some are valiant attempts that may not seem like much but enable us to carry on with integrity. We keep at it daily, doing our part. We are all struggling with the climate monster that has been created. The great news is that we can face it together, in small but strong communities of advocates and activists around the world.

This book includes voices of people like you and me—what inspired us, what we feel passionate about as we face the climate future. The people interviewed are not famous climate activists we all have heard about, although we of course admire those leaders for the paths they have carved out. Instead, the book is about everyday folk who want to contribute in the way they know best. Reading their stories introduces us to a rainbow of actions, each of them only a tiny part of what is possible as we work together.

Participants in the interviews were chosen somewhat arbitrarily. I started talking to a few people involved in climate action whom I knew personally. They recommended others, and those, in turn, more. Everyone had contacts to networks. Writing the book felt like entering an ever-growing circle, connecting with people pursuing a wide variety of interests. What we all had in common was a passionate desire to protect the environment, face the climate crisis, and be guardians of the planet so that future generations could thrive.

The stories gave me hope and renewed energy for climate action. It wasn't that I walked away with a sense that the problem was fixed, or even that it could be fixed someday. Often our climate work is like Sisyphus pushing the rock up the hill, only to see it tumble back down despite our best efforts. Subsidies to protect the climate are canceled; groups dissolve; new disasters strike; there isn't enough support for a new vision. But other times, ideas catch on. We get involved. We plant seeds of change, not knowing the outcome but hoping that what grows will improve the world. We put the seeds in the ground at a favorable place, water them, maybe solicit help for their care, and pray for a harvest we can share with others. The Earth is powerful, always renewing, always active to sustain life. We can be that way too.

And so the voices in this book encourage us never to give up. As we listen to others who are active in their communities, we see glimmers of hope. As climate activists, we are engaged in the most important efforts of our time. We rally with others, pursue our vision for a better world, and enjoy the process of becoming.

How the Book is Structured

Facing the Climate Future Together is a joint effort of many activists. Each chapter features four to six people who agreed to be interviewed. Between the chapters are creative pieces for readers to ponder—songs, poems, artwork, and short plays for street theater or group readings. I wrote these "Interludes" to motivate us for climate action. Together, we join our efforts like members of an orchestra, each playing our instrument the way we know best.

One of my friends compared the book to the very core of jazz: the incredible power of people playing together who happen to be in the same location at the same time. There is no written script for what will happen in the future or what we each must do right now. We create, reacting to the reality in which we find ourselves. We contribute our emotions and talents immediately as we feel inspired.

In the main sections, we hear from the soloists. They bring unique gifts to the jazz band of climate activists. People come with a variety of approaches. We are writers, community activists, organization leaders, farmers, elders and youth, leaders and participants, mourners and enthusiasts. The voices blend to form a coherent piece because we have ideas about how to live right now in ways that keep our planet healthy, and we want to share them because together we are stronger.

Picking up on the ideas presented, after each solo voice there are suggestions for further consideration. The questions raised are intended to trigger personal reflection. They can be used as journal prompts for those who want to think about their own part in the concert of activists. Alternatively, the questions can become starting points for group discussions. What did I learn? What did this solo voice contribute that inspired me? What can I do, alone or as a member of a group, to follow up?

As in a piece of music, there are recurrent themes. The book does not have a plot that unfolds consecutively from beginning to end, leading to some kind of climactic conclusion. Each segment can stand alone; each voice is unique, special, and valuable in its contribution to climate advocacy. It was difficult to divide the book into chapters because most of the participants fit into more than one category: thinking globally and acting locally; reaching out to build community; advocating through writings or storytelling; and coming from a position of deep conviction that the Earth is our home, that we are all interconnected, and that the cause of planetary health is worth fighting for.

A Song of Hope for New Beginnings

"Sun Lights Up the Morning"

"Sun Lights Up the Morning"

Sun lights up the morning; clouds wisp away.
Faith grows that our world will be here yet another day.

With crying hearts and trembling,
all sisters of the Earth
are wondering what might happen
to life and love and birth.
We pray to rise in strength and warmth
as does the blessed sun.
May hope replace despair and fear until the day is done.

Sun lights up the morning; clouds wisp away.
Giving rise to hope and vision for another day.

May we speak with courage for justice and for peace,
Defend creation's beauty, make exploitation cease,
Take steps right now to overcome all violence and greed,
Express our love for planet Earth
each day in word and deed.

Sun lights up the morning; clouds wisp away.
Calling to protect creation on this brand-new day.

We stand near one another, together, arm in arm,
And so, we find protection from hatred and from harm.
And graciously, with tenderness,
we're turning toward the light,
Convinced that love and trust again
will break the force of night.

Sun lights up the morning; clouds wisp away.
Celebrating life and love, our prayer for this day.

Piano Accompaniment

Song by Ingrid Rogers, illustration by Karen Klimpert, and piano accompaniment by Brenda Eberly Bibbee

Part 1

A Circle of Friends

Voices:

Matt Schlegel
George Koerner
Rick Moody
Casey Cameron

Meet the Climate Strikers

It is Friday again. Matt is the first to arrive. He pulls a wagon of supplies onto the large plaza in front of Palo Alto's City Hall and begins distributing a stack of colorful posters near the sidewalk so that passers-by can catch a glimpse of what is about to happen. His Mom, Robin, wipes the wooden round table and takes out snacks. She has been a firm supporter of her son during these last three years of weekly climate strikes. Matt is a professional team builder. He has written a book about teamwork and sees to it that we all contribute in our conversations.

George usually arrives next. He takes the train to our meetings because he believes in public transportation as a way of cutting down on oil consumption. George is in his mid-eighties but still rides a bike and keeps a garden to raise his food. He came up with the idea of conducting climate marches on the first Friday of each month. To reach the most people, he designed a route through the center of town with its busy shops and outdoor cafes, a route that also takes us past banks and law firms that need to be reminded of their contribution to the climate disaster. Palo Alto, after all, is located in the middle of Silicon Valley, home of tech giants Google, Apple, and Facebook/Meta; it is a center with some of the richest people in the country but also a place where many can't afford the exorbitant rent prices and have to live on the streets.

Rick is another one of the regulars. He is well known for his work in gerontology and has published a book, now in its 12th edition, that is used in colleges nationwide. He also crafts a newsletter on climate change that reaches over a thousand people. I call Rick a "collector and connector" because he collects poems, quotations, and cartoons to sprinkle into his weekly email and is always on the lookout to involve more people in his climate work by connecting them to others. Rick likes to be helpful. He stops at Menlo Commons, a 55+ community in Menlo Park, at 11:30 sharp each Friday to pick up Brigitte and me because we no longer drive. We find a parking place near City Hall and join the others with a big smile. Getting together is a highlight of each week for us.

Casey is the one who inadvertently got the book project going. She had been active throughout 2024 supporting the Harris/Waltz campaign, desperately trying to prevent the election of a president who would dismantle climate gains and accelerate oil production. To overcome post-election blues, she thought about working with a group to build a webpage that would include personal stories, videos, and resource lists for climate activists. However, the project didn't work out, and she withdrew from activism for a couple of months to recover from burnout. Casey's idea caught my attention. I have always loved collecting stories, especially through interviewing people to learn from them. So I picked up the idea and ran with it.

But let me take you back to our Friday gatherings. Matt is the initiator who picked up on Greta Thunberg's idea of weekly climate strikes. He doesn't like being referred to as our leader because he strongly believes in shared responsibility and action. Rick keeps teasing him by calling him "our leader" anyway. And it does fit because Matt coordinates the discussions and provides a steady, reliable, permanent witness. There are usually six to eight of us regular attendants. Others come and go. Occasionally people stop to read the signs and join us for the day. Some wave as they pass by or take a quick snack with us.

We begin each meeting with a land acknowledgment. A volunteer reminds us that we are on unceded land that was once inhabited by the Ohlone people. Then we start sharing about the past week. What did we learn? What gives us climate grief? What positive changes are we aware of or have contributed to? There are rules that we have agreed to abide by, like making sure everyone gets a turn and not interrupting or monopolizing the conversation. After the initial round we move to open dialogue about the environmental issues of greatest concern or interest. We are committed to making everyone feel welcome to join us. Effective climate witness depends on being invitational.

So this is our core group at Fridays for Future. My friends nurture me with their wisdom and their commitment. Let's listen to four of them share their ecobiographies. I'll add my own voice afterward in the "Dear Tomorrow" letter.

Matt Schlegel

I started thinking about threats to our environment when I was in college, back in the 1980s, studying science. At that point, I became aware that the American lifestyle created an extraordinarily high carbon footprint compared to the average one in the rest of the world. Physical chemistry helped me understand the scientific part of climate. I realized that we were putting a lot of greenhouse gases into the atmosphere, and it dawned on me that industrialized societies, especially Americans, were emitting much more greenhouse gases than the rest of the global population.

My first reaction was to decide not to have kids, certainly not more than one. I figured if I didn't have a lot of kids, I wouldn't contribute as much to this problem of destroying the planet. I remember actually writing an essay on that when I was in college. It was not that I was scared about the future, or about my kids not having a good chance of surviving the coming crises. I was simply concerned that they would contribute an excessive amount to pollution compared to others.

After I graduated, I started my career and focused less on the climate issue. At that time, scientists came to testify before Congress. We had presidents like George H.W. Bush saying, "We're going to take care of climate." Then came Clinton, and especially Al Gore who ended up being a big champion for climate. So I figured there was at least an awareness at the highest level that we needed to be addressing the issue. And honestly, I became a little bit relaxed, believing that we were on track to take care of it. It wasn't until we headed off to war in Iraq that I started feeling alarmed again. Because war isn't good. Especially wars over oil aren't good for the climate. I knew that we needed to be using less of this stuff, not fighting for more of it.

At that point I became more active, although mostly from an anti-war angle. We couldn't push for climate action much with George W. Bush in office, because essentially his vice president was an oil man. But then, when Obama got elected, I felt like, okay, now we can encourage him to do the right thing.

My activism evolved over time. At the beginning it was more a matter of writing letters to my congressperson or to the president; attending rallies in Washington D.C. and New York City; just making sure that my voice was heard and my presence was felt, at least in the collective. So that's what I did during the Obama years. But I learned something: Trying to make a difference at the national level is extraordinarily difficult.

I remember being puzzled that Obama didn't deny the permit to build the Keystone XL pipeline. He eventually did, but why did it take years? It just didn't make any sense. So I asked my congressperson. At the time I was actually in DC, in her office, and I said. "Why has he not denied the permit for the Keystone XL pipeline?" And she looked at me and said, "Nothing in this town happens without permission from the oil companies."

That was a shock to me, that they had so much influence on everything, even on a Democrat president. But she made it very clear. That's just the way it was. And that's when I decided to pivot to acting more locally rather than focusing on national or even state issues. I would work where I knew I had more impact.

What motivates me as an activist is a deep sense of fairness. Polluting the planet at the current rate is unsustainable. It's essentially borrowing from future generations—affecting not just human life but all life. Everyone deserves to have a qualitatively safe and thriving environment. Not just those who can buy this somehow. I feel that we are essentially stealing the future of our kids and of all life on the planet. That seems deeply unfair to me.

I've always been pretty attentive to nature. I have done a lot of camping and hiking. I was in the Boy Scouts. When I was younger, I lived in Pleasanton, California. Back then it was much more rural than now, surrounded by fields and creeks. I used to play in the fields, make tunnels through the grass, and play in the creek. I was very much connected with nature throughout my childhood. Interestingly, climate change was already noticeably happening even back then. My grandfather had a place in Pleasanton with a stand of pine trees. One year all the pine trees died because of the pine bark beetle. They had come north when the winters stopped being as cold as they once were, cold enough to kill the larvae. So the larvae of the pine bark beetle survived and devastated the trees the next year. I saw that in the 70s. By now they have become a huge problem all over the state, especially in the Sierras.

When I think about young people today, I worry about the effects of climate change they will have to live with. But I have a message for them. First of all I want to tell them that there are adults who really care about what is happening and are eager to take action. So, if they are becoming concerned and care enough to want to do something, then they can find communities of people of all ages to collaborate with.

Once you understand the scope and magnitude of the problem, it can be overwhelming. It's huge, and we all need to be doing everything all at once. But here is the thing: if you want to take action, focus on the aspect of climate change you're most passionate about. If you are really passionate about it, then you will persist and you'll keep working on it. That's what we need: Be persistent. And do something that you love doing—not out of a sense of obligation, but with enthusiasm. Work on what you feel deeply about, whether that's restoring wetlands or cleaning up ocean environments or taking care of an endangered species or helping your community reduce its carbon emissions or making the community more resilient to climate change—whatever it is that you are particularly passionate about. Do that because together we need to do everything. Trust that there are other people out there who are equally passionate about those other things, and who will act and advocate for them.

Before the pandemic, I worked for three years on a book about team problem-solving. It wasn't specific to the climate, but the subtext was there. I kept thinking, what can I contribute in my work to helping humanity come together and find solutions for the common good? How can we apply these principles to address the climate crisis? When I finished the book in 2020, I decided I really needed to put into practice what I wrote about, to start taking a more active role, not just an intellectual one.

I had already been thinking that I wanted to work locally. One of the people who inspired me was Greta Thunberg and her activism through climate strikes. I liked the pace of standing up for climate on a weekly basis. Monthly meetings allow you to put concerns out of your mind, perhaps think about them just before the event and maybe a little bit afterward—but then you have three weeks to concentrate on other matters. I wanted to keep the climate issue on the table more consistently. That's why I gravitated toward doing a weekly climate strike. Also, it was important to me to build a community that felt that same way. And so I've been doing that. This Friday will be the 157th consecutive week of meeting in front of Palo Alto City Hall to think with a group of people about climate and about what we can do to help address the problem.

I started meeting with people who had similar concerns. I live in such a great community with lots of action and involvement options. I was supporting their initiatives and soon found a couple of people who liked the idea of meeting weekly. One such group already existed. They called themselves the Climate Mob. They were meeting at somebody's house, but they liked the idea of becoming more public to get the message out. And so a few of them decided to gather in front of City Hall. That's how it started.

It is important to be persistent. When you're the only one who shows up, that can be disheartening. A few times, I was the only one out there. I'd put out my signs, take out my guitar, and just play music and write climate songs. I kept going every week, and now there are always people who stop by or who come on a regular basis.

Over the course of the weeks, I had made different climate signs with messages to spread out in front of City Hall. At the end of the meeting, I'd have whoever was attending grab a sign to hold up, and then we'd take a picture. That way I'd have a photo to promote climate action and also talk about it in a weekly newsletter.

I started the newsletter because it gives people a sense of what we have accomplished, and it also invites people to participate in the future if they want to. I try to highlight what people are working on and what we set out to do, whether it's a march or singing songs or starting a new project. We celebrate climate successes. I always ask people what's on their mind with climate and everybody reports different things, whether it's a concern or a cause for celebration. I want to capture all the actions that people are taking or issues that they want to promote. Those are the general themes of the newsletter. And also I want to give that sense of continuity that, yes, we're still meeting every single week. People have busy lives, and things happen, but I want them to know that we are there and they're always welcome to come and join our community.

I love it when we collectively come up with an idea and rally around it. In the beginning we promoted getting people to replace their gas water heaters. We got the city to create a program and help subsidize heat pump water heaters. Right now, that program is in place. You can essentially replace your gas water heater with an electric heat pump water heater for the same price. So that climate initiative was a huge accomplishment.

More recently, about a year ago when we were brainstorming, we came up with the idea to do a monthly climate march through downtown. We're taking advantage of the fact that since the pandemic, a lot of restaurants have opened street seating. At lunchtime, a lot of people just sit around or stand waiting for a seat or move on the sidewalks. We can bring attention to the climate crisis to all these diners outdoors. The Climate March has been so successful that we have continued holding one on the first Friday of every month.

When I imagine a better climate world in the coming decades, the first thing that comes to mind is community. Networking and supporting each other is crucial to help us navigate the crisis.

Some aspects of the climate future look really grim. We inaugurated a President who is a dedicated climate denier, committed to drilling as much oil and fossil fuels as can be done during his term. That's a dark prospect. What gives me hope is that the rest of the world does take climate seriously, and many Americans do too. They are moving forward with climate action, in spite of what is coming out of the mouth of the President. Take, for example, my own state of California, where our governor has already announced that we will start to take precautions now to make sure that these measures we've taken can't be undone during the current administration.

I follow a lot of people on YouTube and social media who are concerned about climate. They talk about the crises and about what they're doing. It makes me happy how resilient they seem to be, even in the face of the new administration. Everybody knows the announced policies are not good for the climate. But you know, we're going to just keep working to address the crisis constructively. We're not going to give up. And that's what gives me hope. Even when things look dark, people can be resilient.

Working on climate is very challenging and, at least for me, emotionally fraught. In my climate activism, I allow myself to feel all these emotions. I want to feel them because I know that a lot of people feel them too. And I want to develop empathy—deep empathy for what people will likely have to go through if they aren't going through it already. At first it was challenging for me to allow myself to do that. I am blessed in that I normally don't dwell in emotions too much. But I thought it was important to do that. And so I did it. The result was that I shifted my emotional state. I'm a very different person now.

When I got deeply involved with climate and I felt such pain and grief, I coped by picking up an instrument. I learned to play bass and I joined a musical group. That was very helpful for me because music is a great way to process emotions. It's cathartic and it's also a purposeful way to engage with a community of people. My music-making friends have always been sympathetic to my climate concerns, even though they aren't climate activists. I'm very grateful for that. So I have these two distinct communities—my band and my climate group.

The emotional journey of putting myself through the whole grief process was important for me. Now I am at the point of acceptance. Acceptance for all we have lost. All we're losing. And all we're going to lose. You get to that point of a post-tragic state where you realize that one molecule of greenhouse gas that you *don't* omit is helpful. You want just to keep focused on making things as good as they can be. That action might not be helpful to you, but it will be helpful to some life. You're going to make the planet just that much better. One molecule at a time.

For Further Reflection

- How do you react to the unfairness that people in the global north have a much higher carbon footprint than those living in the rest of the world?
- How can we best hold oil companies to account?
- Have you observed changes in nature due to climate change?
- Do you agree that we are stealing the future of our children?
- What, of all we have already lost, causes you the greatest sorrow?
- How do you deal with climate grief?
- Share an example of collective action. What are the advantages of working in a group?
- Give examples of each of the following: a) working in a group to bring about positive change, b) being resilient, and c) not giving up.

George Koerner

I enjoy our Fridays for Future meetings. Getting together with others who have similar concerns is very reassuring. We are more like a support group, which is really needed these days. The group gives me something to look forward to each week.

I get to the meeting by bus, so I don't have to drive at all. That's important to me because I am interested in climate and the environment. It's difficult to say when I first noticed that something was wrong with the climate. One instance I remember is touring America on a motorcycle, going up to the Columbian Glacier and seeing how far it had melted back. That was in the 1970s. I went back five years later to the same glacier, and it had moved way up the hillside. And I said to myself, "Why aren't people panicking? This is not right!" As far as I could tell, glaciers all around the world were receding,

Another clue I had about environmental degradation was the smog in Los Angeles. It's really hard to pin it down to one specific thing that upset me because the damage is cumulative. It just keeps going. In LA people talked about smog. Of course, we also listened to the Tom Lehrer song, "Pollution." The lyrics told us, "Don't drink the water and don't breathe the air." He was singing that back in the late 50s. I was born in 1943, so I was a teenager at the time.

I loved the outdoors. My favorite weather during childhood was fog because it felt fresh and was silent. It muffled the sound, and it kept people indoors. I liked that a lot. I grew up equidistant between Golden Gate Park and San Francisco Zoo. Both were free for visitors back then, so I spent a lot of time there. We could take Sunset Boulevard and I'd get on my bike and go to either one. Occasionally my family went to the beach, which was just within a mile or two of our house. So I grew up near the ocean in the fog.

For vacation, my parents went to Camp Mather up in the Sierras. It wasn't very often, but we'd go up there for a week and sleep in a cabin with no heating and no insulation, just more like a wooden tent. I really enjoyed being in the mountains.

I remember a day in San Francisco, I'd say mid-80s, when I was standing near the Cliff House looking out to the Golden Gate Bridge and I saw an orange cloud coming out from under the Bridge on the north side. A big orange cloud was drifting out toward the ocean. Normally we couldn't see much because of the fog, but here was this orange cloud moving like smoke, floating out there. Nobody else seemed to notice it or find it startling.

I never did find out exactly what caused the cloud. I had no idea where to find information on that. Mostly, I guess I became aware that other people weren't upset about the changes that were going on. My first hint at pollution and its consequences was back in the 1950s. In several murder mysteries on TV, people died from asphyxiation because they sat in a garage with a car motor running. And I got to thinking, well, if you can get killed in the garage with the car motor running, how long would it take for that car to run before the Earth's atmosphere is polluted enough to kill you? I was aware that the Earth's atmosphere was finite, and I kept asking, "How many cars do we need to have running for resulting pollution to affect and bother you?" When I mentioned this to people, they looked at me like I was from another planet.

I did drive a car when I was young. I owned a 1962 Volkswagen bus. In fact, I still own it, although it hasn't run in years. It was very good on gasoline and could carry a lot of gear. It got 30 miles to the gallon, very efficient for the 1960s. Now it's sitting in the garage but we'll soon sell it. We have two buyers eagerly waiting for it.

I stopped driving at age 81 and didn't renew my driver's license. My wife still has a car, but we avoid using it as much as possible. Last year the two of us drove only 850 miles total. Mostly we go by bus and bicycle. I don't use the bike as much as I used to, and neither does she since she broke her leg. Now we're mostly walking or using public transit. The only places we visit are those we can reach by bus or public transportation. Last summer we spent our vacation in Santa Barbara and San Diego without driving at all. In San Diego, we got a senior bus pass, figured out the bus lines, and easily got around to see the botanic gardens and the pandas and all that.

Choosing public transportation is one piece of my attempt to help the climate. Another choice we made is to eat healthy organic food to avoid the poisons in much of the processed food. We try to keep the water clean, avoiding the chemicals we've been conned into using, like Roundup and things like that in the old days. Everybody was told, "Oh, this is safe. You can use it to get rid of weeds." No! We are exposed to a constant inundation of things that are really bad for us but are profitable for companies.

Caring for the planet takes persistence. Just get out there, talk about it. Currently, I feel barely able to carry on because what's going on is so horrendous. What we are doing politically and ecologically, it's all one destructive package. I feel like I'm watching a train wreck in slow motion. Even trying to speak to people about it is frustrating.

Some people say that they are interested in saving the planet, but then they act in ways that accelerate the climate crisis. I was involved in Extinction Rebellion for a while. The group decided they would all fly to a protest back east. How do you protest against an oil pipeline and use a plane to get there?

We are not fighting the climate crisis. We ARE the climate crisis. We are creating it by building up CO2 in the atmosphere. So my motto is, "Do not feed the dragon."

The dragon thrives on CO2. Every time you release CO2, you feed the dragon

We have made the necessary changes in our home, like installing a heat exchange water heater. We've had electric heating in our house since we bought it 40 years ago. The house has a 1904 fireplace that we rarely burn fuel in, but if the house really gets cold, we can use firewood to heat the living space in an emergency. The only gas appliance we have left in the house is a 1920s gas stove in our kitchen.

Outside, we have three compost bins that get rotated. And we have a wild garden space with a valley oak that's probably 250 to 300 years old. It's shared with the home of a neighbor, covering both properties. We planted a red oak nearer to the house. The two trees cover the whole backyard during the summer, so we don't have to worry about cooling because those trees shade us very well. And we have understory trees like a dogwood and a row of yews. We have an abundance of wildlife—butterflies, squirrels, gophers, and lots of birds.

I've been trying to preserve the Earth's resources for most of my life. In the early 70s I started riding my bicycle to work every day. Even toward the end of my employment, I rode ten miles each way. People looked at me as nothing more than a weirdo, but I kept trying to tell them, you know that going by car is not good for you or the environment.

One thing I learned as a young person was that plants absorb what we exhale and return oxygen to us. There is reciprocity. But when you burn oil, how does it enter the environment to go back to be recycled again? And of course, it doesn't. Once we burn it, it becomes CO2. It doesn't go back for millions of years, if ever. That's very scary to me.

My wife and I met because we both bicycled to work. I noticed she also rode a motorcycle. And at the time I had a motorcycle, and we used to tour across the country by motorcycle rather than drive a car or fly. So we had that in common. Once I was able to, I bought a house three blocks from my workplace, so I didn't have to do the 10-mile bicycle ride but could walk, even go home for lunch. My wife and I share the same values. We like walking and hiking a lot. She loves nature also. She's the one that feeds the birds, buying 25-pound bags of bird seed.

I find it offensive for humans to do damage to the planet. We are all responsible for our actions, and I try to do as little damage as possible to get by in life. For me, it's not so much a matter of fighting climate change; just stop causing it. Whatever I do, I want to do the least harm while I'm here.

No one explicitly taught me that disregarding other life forms is a harmful thing to do. I have no idea where exactly that awareness came from. It developed cumulatively over the course of my life. Thanks to my grandmother, I got to travel as a kid. In the 1960s, she wanted to go back to Germany to visit her family, but she couldn't find any friends that would go with her. So she took my sister and me. We went to England first and then to Germany and Switzerland, all without flying a single time. We took the train across the country, took a ship across from New York to Southampton, and purchased a microbus. We drove the bus on the Continent and eventually shipped the car home, went by ship back to New York, and took the train across the Northern route home again. I think we were gone for three months altogether. It was an amazing experience.

These days, our Fridays for Future group reminds me that there are people who do care about what happens to our planet. Matt does an amazing job. He's a great catalyst for the group. He gets stuff done and works very hard at it. Rick's weekly newsletter is reassuring to me too. It's an honor to be part of the group. We get together and talk, sometimes about more than just climate. Sitting outdoors gives us visibility, a message for people who walk by.

I was inadvertently responsible for starting our climate marches. I noticed that during the pandemic, many restaurants in the center of town had set up tables outdoors in an area that used to be parking strips. Now you could actually walk down a public space. In essence, you're walking through the middle of a restaurant, and they can't stop you. So I suggested that we use that opportunity to pass by with our signs. Of course, we have to do this respectfully so as not to interrupt people's lunch. We don't make noise while going through that area but just plod through and are seen. It's kind of like watering a lawn. You can't pour a year's worth of water on a lawn all at once. You have to apply it gradually and let it soak in. My intent for the monthly demonstrations was to proceed gently, so we can at least get some attention but not overstep the bounds by disrupting people's lives.

Sometimes talking to people can move them a little in their way of thinking about climate. I have been part of a book club for at least seven years now. At first the members thought my wife and I were a little weird for taking the bus everywhere and pushing train travel over airplane travel. By now, three or four members of the group only take the train and not fly. So what we shared with them bore fruit in the end. They saw the problem and made changes.

When you envision the future, given the human animal and our history, I am not very hopeful about where we are heading. Often I think it can't get any worse than this, but then it does. I feel really frustrated. It's kind of like when you're teaching someone to drive and you say, "Look out for the brick wall! There's a brick wall around the corner!" But instead of paying attention, they accelerate and hit the brick wall.

It's humans in general, all of us. I worked in computers for years and saw how we would attack a problem by treating the symptom without identifying the source. Many times we'd find, "Oh, this number's wrong.

Correct that number and everything will be fine." But unless you burrow down into the logic to find why it happened, you're not solving the problem.

Here is an example regarding Earth care. People engage in greenwashing, saying "We'll have this new thing that's going to get the carbon out of the air. We're going to recapture carbon, like a perpetual motion machine." It sounds good, but when you get down to the nitty gritty and how it actually works, it's not very likely to work. They haven't assessed the whole problem. We keep producing carbon, and nothing will change until we stop.

Do not feed the dragon. If you're building a bonfire and the bonfire is getting out of control and your house is catching fire, you don't go out and get more firemen to run in. The first thing you do is stop feeding the fire.

Basically, that's what I see. People feed the dragon without realizing it. Deep down, I feel that there is no profound change in sight. I want to believe in the possibility of a change in human behavior, but we may not survive long enough to get there. It's a really close race.

If anything gives me hope, it's people taking creative initiatives. Fridays for Future. The Green Party. Extinction Rebellion. Greta Thunberg. Rick, Matt, and others from our climate action group. Efforts like that are going on all over the world. People keep going, even though they're ignored or the powerful keep trying to squelch them. But we keep trying. We talk, read amazing books, learn, attend meetings, and take local action. We keep trying. So many people are desperate to get this problem solved. It is disheartening that we seem not to be as effective or fast as we need to be.

For Further Reflection

- Is your concern for the climate reflected in your transportation choices?
- Do you agree with George that we are the climate crisis?
- Give examples of how we disregard the value of other life forms.
- How can we protest in the streets in ways that are effective, respectful, and invitational?
- Have you ever moved someone else to take climate action?
- Comment on George's warning, "Do not feed the dragon."

Rick Moody

I started a weekly newsletter about two years ago. At first I wanted to call it "Leaving a Legacy," because it was intended to address both the Social Security program and climate change; but then I realized that people usually don't see the two topics as closely connected. However, they both challenge us to think about the future. Leaving a legacy is important. That's really what climate action is all about. Aging and climate are related in that both make human beings encounter their limits. We become aware of finitude, as the Existentialists call it. And we react with fear or with denial.

In my newsletter about climate change, I include poems, quotes, and even cartoons. Why do I begin my newsletter with humor? One reason is that I enjoy laughter. Another is that maybe otherwise people won't even bother reading the rest. Most importantly, humor is important to break the numbing effect of fear and denial.

Six months ago, before my climate book was published, my editor said, "You've got a great book here. There are just two problems with it." I said, "Really? How so?" He replied, "Well, it deals with climate, and that's a downer. People aren't happy about that. In addition, it deals with aging. People are against aging. They definitely don't want to age!"

Two strikes against you, three strikes and you're out. Well, the book is out. Like the newsletter, it's called *Climate Change in an Aging Society*. I continue to publish the weekly newsletter, but not everybody wants to get the newsletter. Not everybody wants to read the book. That's okay. My job is to help people who want to do something to leave a legacy. And that's what I do.

I guess my strength is that as an elder, I'm willing to say things that other people need to hear. There are great artists throughout history who become less restrained as they get older. They're willing to say what they think or to adopt a style in the art world that's more daring. The Germans call it "späte Freiheit," meaning "late freedom." I think late freedom is another name for wisdom.

It's not that every artist is wise. The great art critic Ananda Coomaraswamy, who served as curator of Indian art at the Museum of Fine Arts in Boston, said, "It's not that the artist is a special kind of person. It's that each person is a special kind of artist." That's another way of saying that each of us has challenges and gifts. The Jungians call it individuation. I call it becoming the person you were meant to be. That's what late freedom is: becoming the person you were meant to be, able to say the things that you need to say, and maybe people need to hear. And I think that's true for climate change as well as other causes. So I don't claim to be wise, but I do claim to have more freedom than I had when I worked for the AARP.

People should think globally but act locally. I send the newsletter to 1,200 subscribers all around the country. Many of them write back to me. I try to connect them with people they may not know about, either nationally or locally in the town where they live.

I'll give you one example of this. Peter Whitehouse, an old friend of mine, is a professor of neurology at Case Western Reserve University Medical School in Cleveland. Some years ago, the river in Cleveland caught fire because of discarded petrochemicals. Like other places in our country, Cleveland had deposits of residues from fossil fuels that endangered the community. Peter Whitehouse recognized this as a huge problem. He dressed up as a tree and said, "I can't continue to be a medical doctor anymore. I have to become a tree doctor. We can learn things from trees. Trees communicate with each other and help each other. We need to do that as human beings too."

A woman in Canada near Vancouver who did research on trees discovered how trees help each other—and not just other trees but even fungi. All these different life systems are connected with each other. When we destroy a forest, we're destroying more than just one system. We have to be very careful about cutting down trees. So what Peter did was he brought together elders like himself and said, "We need to hear from you about what you have done in the past to protect Cleveland and the area around it." And they were happy to do that.

Peter and his wife created an intergenerational school to work together with young and old people. Professionally, he had become a world-renowned expert on Alzheimer's disease, but in retirement he turned his attention to something different. Alzheimer's is a cognitive brain disorder that causes cognitive decline. We forget what we need to remember. That's part of the problem with climate too. We've forgotten how the world ought to be, and we've taken it for granted. What Peter has done right there in Cleveland is to bring together different generations so that the young can learn the lessons from the old and can carry on the message.

Whatever our age today, it's true that fires, floods, heat waves, and drought are in our future. Some people are now waking up from the climate nightmare. Others are still sleeping or hiding in denial. There's an old joke, denial (the Nile) is not just the name of a river in Egypt. It's the name for reluctance to face what we already know. That's the reason why I'm interested in dream work, including climate dreams, by the way, because dreams show us what we already know but cannot yet see. That's true about climate as well. People know about climate change, but they don't see the results or implications yet. Eventually they begin to see it in the form of tangible extreme weather events, like fires, floods, droughts, and heat waves. Those four are what I call the "Four Horsemen of the Climate Apocalypse."

We've already experienced terrible climate-related events. For example, we had devastating fires in the Los Angeles area recently. The fires were not limited to just one place because that's what happens with fires. They leap even over highways. And that's why we can't isolate one place from another.

That brings up an important message about climate change: We can't live in isolation from one another, from nature, and from other species. People who live in denial may say, "It'll never happen to me." Or they'll say, "It's all over anyway, it's too late." Both attitudes result in the same response, namely passivity.

So the question is, what can activate people? We have to help people take the next step and show them what they can do, but we have to take it one step at a time. First we must ask ourselves, "What may galvanize them to say, yes, we can do something"? It's important to pay attention not just to the threat but also to create opportunities and show things that succeed.

Periodically, I will put little articles in my newsletter about actions we can take individually or about what people in other parts of the world have done that we may not have heard about. Bangladesh, for instance, is a poor country right in the center of climate threats like global warming and ocean rise. The people there depend upon the runoff from the rivers of the Himalayas, and much of the population also lives close to the ocean. Years ago, they realized that they needed to take action to mitigate climate disasters. They figured they could use their schools and other buildings as places of refuge when floods come. So they started building more structures they could rely on for protection. Here's a great example of how people in the rich global north can learn about mitigation from poorer countries. We can learn from people who are vulnerable because they've had to face up to the dangers of climate change. Soon we are going to face up to that too.

I started paying attention to climate change when I was about 30 years old, but then I stopped because I was busy with my life—having a family, writing books, being an academic. Now I am 80 and have granddaughters aged eight, five, and 10 months. I won't be around even by the time they go to college. What do they need to hear from me? I wrote them a letter expressing that I felt guilty about not having paid enough attention to climate change until recently. But I also told them that we older people did the best we could—not everybody, true, but plenty of people had tried to warn others and tried to live a life that protected the Earth on which we live.

When I update the letter, because I do from time to time, I include some practical things that I did in my own life. I don't do it to praise or justify myself. I simply want readers to hear that there are actions they can take and lessons they can apply in their lives. In the future, it won't be just a few people who have to do these things. Many more will need to step up because the threats will be everywhere. The oceans are rising. Even if we were to do everything needed to reduce greenhouse gas emissions and stop using fossil fuels, even if we did all of that today, right now, it wouldn't make any difference in the sense that the oceans will still keep rising. The heat is already out there, melting the ice caps of Greenland and Antarctica.

That doesn't mean it's too late to do anything. No, it's not too late, but it means that we need to recognize that people in the future will face a different world. In many respects, it will be a worse and more threatened world than the one I grew up in. So we elders are trying to communicate what we have lost and what we will face and what we can do. Bill McKibben is a great example of somebody who sends out that message. He is a great model for us.

What I want to say to the youth of this world is the same as to the elders: It's not too late. The youth sometimes feel they've been dealt a bad deal. It's not fair that they have to live with the consequences of what their elders left behind. There's truth in that, and that's the reason why elders have to be responsible. Rabbi Joshua Heschel used to say, "In a free society, some are guilty, but all are responsible." So that's what the young people need to hear. Yes, some are guilty, but now, the young people are responsible too. Now we're all in this together.

Some solutions may lie with technology, but technological advances are difficult to predict. A great British physicist, Lord Kelvin, said around the year 1900, "Radioactivity is a hoax. There's no such thing as radioactivity." Kelvin was so prominent that they even named the temperature system after him, "Kelvin degrees." But he got it wrong. Marie Curie was already starting to figure out what radioactivity was, and she proved that it was true. So much for predicting advances in technology. We can't predict the future of science.

We can't predict future technological solutions, but we can predict some of what is to come by looking at the past. The cost of renewable energy has fallen drastically. The price for solar panels has come down by 80 or 90% in the last 10 years. Could it come down more? Yes. But even if it doesn't, we can make those solar panels more available, more efficient, more recyclable. Battery storage has made some great gains. We can create more efficient batteries.

Some people say, well, maybe we could have fusion energy. Then again, that type of technology is 10 years in the future because it means finding ways to bring hydrogen atoms together to make helium. That process goes on in the sun. If we could replicate that process on Earth, we would have a seemingly limitless source of energy. It's a great idea, but it hasn't happened. The same is true for some plans involving geoengineering or carbon capture. They look promising, but they may have a downside. My problem with all of these proposals is that they could distract our attention from what we need to do right now.

Climate organizations can have a huge impact. "Third Act" was founded just a few years ago by Bill McKibben and now has something like 70,000 members active in chapters all around the United States. Rebecca Solman, who collaborates closely with McKibben, wrote a book, *Hope in the Dark*. I love the title. In the book, she writes about people in Eastern Europe who lived in places "behind the Iron Curtain," people who knew there had to be a better alternative. And eventually there was. I think that that will happen with climate change too. That's why I have hope. I don't believe that we will always remain in the dark.

When some people hear the word "hope," they think hope is just an illusion. Well, it can be an illusion, but it doesn't have to be. The great environmentalist David Orr once said, "Hope is a verb with its sleeves rolled up." In other words, hope means action. It means you do what you can do. You don't just complain. You don't just send social media messages. There's nothing wrong with social media or with complaining occasionally, but what really counts is whether or not you act. What you do is what carries the message.

The climate crisis has already arrived. Whatever our age today, it's true that more fires, floods, heat waves, and droughts are in our future. We do have a choice about what to feel and what to do about it. Instead of falling into denial or despair, we can find opportunities to act, and action is the meaning of hope. Instead of giving in to learned helplessness, each of us can engage in ways that make a difference.

For Further Reflection

- What do you want your descendants to know about your actions regarding climate change? What personal legacy would you like to leave?
- Can jokes or cartoons be effective wake-up calls to draw attention to a problem? Give an example.
- Do you know of elders who experienced "late freedom"? What did they do?
- How can we be more like trees, upholding each other?
- Which of the "Four Horsemen of the Climate Apocalypse" scares you the most, and why?
- Do you know about mitigation efforts in your community?
- Which technological advances give you hope for our climate future?

Casey Cameron

I didn't have models for activism early in life. My parents weren't politically involved. They were Democrats, but we didn't talk much about political issues. However, my father was a devoted organic gardener. When I was a child, I saw his reverence for nature. He told my brother and me about composting and growing things. Frankly, as kids we were not all that thrilled about gardening, but we loved him and saw his enthusiasm.

Most of my life, I stayed relatively unengaged regarding climate. As a single mom, I threw myself into employment to protect my daughter by making enough money to raise her properly. All I could think about was, "I can't get laid off. I just have to put everything into these jobs." I first worked in a high-tech firm in Boston as a technical writer. I got a decent job with good pay by joining the burgeoning computer industry. I worked with a group of very well-regarded engineers.

The company offered us all great jobs in California. When I got here, I reconnected with one of my oldest friends whom I'd known since eighth grade. She was a teacher in Berkeley and a lifelong activist. At first, I saw her as someone completely different from me because her orientation was all political. I mean, everything she did, everything she thought about had to do with activism for various causes. She had participated in the anti-nuke movement, but then she got into circles of activists who were starting to talk about the environmental crisis. Eventually, climate action became her central focus.

I distinctly remember sitting in her backyard on a beautiful sunny day, listening as she talked about her climate work. I was nodding along because I very much respected her, but she knew I hadn't done anything to address the crisis. All of a sudden, she stopped and said, "Hey, you are listening to me and you understand the problems we are facing. So how come you're not doing anything?"

It was kind of a dig, but it was great. There are moments when something registers in your mind. I suddenly realized that yes, I did care about what was happening to our planet. My daughter had gotten older by this point and was much more independent. I still had a big corporate job, but I knew I should get involved. My friend showed me a whole curriculum she had developed to talk to people about climate issues. She had designed these huge, very engaging, and convincing placards with pictures of pollution and graphs showing how global warming impacted life on Earth. She and a couple of her friends took that information to all the senior centers in the area. I was profoundly impressed with her initiative.

I learned a lot more from her after that. She inspired me with her zeal and her frank disdain for my inaction. She recommended books to read, and she told me about a climate group I could join. So that's how I got involved. I started reading more about the climate crisis on my own and began talking to people about it. Fast-forward to today, and there is no issue I care about more than defending the climate and our precious ecosystem.

My husband, who's a voracious reader, also knew a lot about global warming. When I brought up the subject, he was like, "Oh, yeah, this is a real problem. It has been a problem for years. We just didn't hear about it much because the big corporations have tried to sweep it under the rug." My husband is as upset about environmental degradation as I am. He is introverted and doesn't participate in climate action, but he approves of what I do and supports me.

The best part of environmental activism for me is working with others who care about the planet. To me, that's the joy of my life. I've been involved in climate action long enough that I know quite a few people in the movement.

When I first started, I joined a group called "Silicon Valley Climate Action Now." It doesn't exist anymore. We had certain pet projects, like moving our money out of the bad banks and protesting. Sometimes it got a bit scary. Every week, a group of us would go into a bank during business hours and disrupt proceedings. We had posters that told about how the big banks are funding the oil and gas industry. We got a lot of negative pushback during our protests. In one bank, a couple of customers who were in line to get to the tellers turned around and yelled at us, "Get out of here. You have no right to do this." It was confrontational and disdainful.

I wouldn't participate in this kind of protest anymore today. It didn't convince anyone to change their behavior. I ended up leaving the group because the next thing they planned was to mount a big action that would have required a lot of driving. They held lots of meetings. They decided to get a whole bunch of people with kayaks to go into the San Francisco Bay for protests similar to those that Greenpeace organizes. It was more like direct action, and I didn't want to do that. Frankly, I was scared.

The next group I joined offered classes. One was on holding difficult climate conversations, like how to conduct yourself so that you don't turn people off. Another one that I joined was about how to move your money out of the bad banks. It was several weeks long. Each week we were encouraged to take a specific step. It's complicated to change banks for a lot of people, especially if they have mortgages. I went to meeting after meeting and eventually pulled out of two banks, Wells Fargo and Chase. Now I use a credit union for my checking and Visa card.

The groups I joined next met mostly on Zoom, but members also knew each other personally and would show up at climate events. We continued to protest at banks, but without being disruptive. We just stood outside legally with our signs.

I also joined Elders Climate Action, and I still support that organization. I love their newsletter, and I follow up with actions they recommend. Then during the pandemic, I really doubled down and started using an app called "Climate Action Now." It's mostly for sending emails via handheld devices to corporations and to our representatives. I mail about ten of those each day. To test the effectiveness, I once asked the guy who started the app, "How do you know this works? This seems almost too easy." And he said, "Well, you know, we did a lot of research. It's a numbers game. Organizations and politicians have to prioritize. So if you send a message 'Hey, Coca-Cola, you said you would get rid of plastic,' and if enough people do that, they hear you. It's just pressure, pressure, pressure." So I continued doing that.

And then I found Matt and his Fridays for Future group. I feel sustained, and I like those who come. The people part of every action is always important to me. It's what keeps me going.

Prior to the election, I got involved with Elders for Sound Democracy. I wrote their newsletter. For about four months, I worked with a fellow from Oregon on that. We were great partners. We would have meetings almost every day talking about what should go in the newsletter, how we should address the issue, and who might contribute. The newsletter was very successful. We got excellent feedback. People really liked it.

I'm a communicator, so I like talking to people in a way that helps wake them up to the climate situation. You can't approach the topic in a didactic way by giving them a tedious lecture. I use opportunities as they arise and start little conversations. For example, last summer I was getting my car smogged and there was a guy behind the desk on an extremely hot day. I was filling out the paperwork when he said, "Boy, it sure is hot enough." And I replied, "Yeah, well, you know, climate change is real." He looked at me. "Oh, right," he said dismissively and added, "I don't believe in that." But that got a little dialog going. I made sure not to sound "holier than thou." I just said, "Oh, you should read a bit about it because it's important."

One time a friend in New York wrote in an email that he and his partner were anticipating having to drive home in a huge storm. I replied, "I hope you'll be safe. You probably know that the warming ocean (from carbon emissions) is stirring up more and more unpredictable weather. Scientists are saying that we talk about the climate problem as something that can be mitigated, not as an inevitability."

Just practicing dialogues is important. I often allude to climate problems at the grocery store. At the checkout counter, they always want to put my produce into plastic bags, and I say, "No, thank you. Plastic bags are out now. Didn't you know that? Nobody wants plastic bags." I sort of joke around with people and keep the conversation light, but I find little ways to introduce ideas to protect the environment.

Just flinging facts at people, particularly if the facts are at all negative, doesn't seem to work. I try to appeal to the best qualities in people when I talk to them. One time I said to an older man, "I know you to be a person who cares about animals. Are you aware that 50% of animal species have died or are in danger of disappearing?" I'll give people a little fact to get their attention. I start with their own interest to show that we share the same concern.

My daughter has become an activist, but not around climate. I've been working on getting her more interested, sending her information to call her attention to climate developments. Everyone can provide tiny bits of education like that. The more people you talk to, the easier it becomes. Let people know that climate is not something you just dismiss. Climate change is big. We have to be aware of what is happening.

Sometimes my efforts don't succeed. I tried to start a little climate group in my townhome complex. People were sitting outside in the nice weather, and I asked several, "Are you interested in climate change?" They said, "Yeah, I'm very concerned." So I asked, "Would you like to get together for a discussion group?" But then they said they were too busy. I didn't get anybody to join. One person seemed interested, but when I sent her an email, she didn't respond.

When people are reluctant to get involved, it's discouraging, but I tell myself that I'm fighting for our future. It's an important overall goal. Also, I can handle disappointments because I'm fortified by my own spiritual life as a Buddhist practitioner. I meditate, and I feel good about what I'm doing. It will be a long, slow battle. Whether it gets won in my lifetime, I don't know. The people that I interact with talk a lot about "engaged Buddhism" where you are not just sitting on your cushion. You're doing things to help make people's lives better. My spirituality is a strengthening force in my life.

One initiative I am inspired by is the "Our Children's Trust." It started with young people acting as plaintiffs in court saying they had a right to clean air and clean water. Now lawsuits have been filed all over the country. I donate to "Our Children's Trust" because the organization needs funds to pay the lawyers who fight their cases. They've won two big cases now, in Montana and Hawaii. That's exciting.

To me, the climate crisis feels like a huge gathering storm. The ocean warming is tragic. I think about the multitude of effects: the destruction of wildlife, the ruin of ecosystems that have sustained life on Earth for millions of years, the Gulf Stream becoming static, and the suffering of animals who lose their habitats. Several of my friends on the East Coast have told me about the so-called "red tide," those huge algae blooms that turn the water red. The algae are a terrible threat to the creatures who live there. My friends say they can't stand going to the beach anymore because it's so awful. I also worry about kids growing up and having to deal with the climate disaster.

I'm by nature an optimist, so I stay hopeful, although it's daunting to come up against this dark force in the world that is destroying our ecosystems. I believe we have to keep fighting. I know it's the right thing to do.

For Further Reflection

- During your childhood, did a family member or friend guide you toward an appreciation of nature?
- If you had witnessed the scene at the bank that Casey described, how would you have reacted?
- Where would you draw the line at a protest event? What scares you? What actions would you not participate in?
- What influenced your choices of where to do your banking? Are the banks you use eco-friendly?
- Casey describes circumstances where she injected climate information into the conversation. What do you think of her approach? Have you taken similar initiatives?
- How do you handle the disappointment when one of your efforts doesn't produce the result you had hoped for?

A Letter for the Grandchildren

The Dear Tomorrow organization suggests writing a letter for our descendants explaining what we did to keep this world safe for their future. I want to share mine with you below and encourage you to compose your own.

Dear Margaret and Caroline,

I picture you reading this around the year 2050, long after your grandpa and I are gone, wanting to know whether we had a part in fighting for our Earth.

Growing up in Germany, I often wondered what my parents had done to oppose the Nazi regime in the 1930s that led to WWII. I was so disappointed when I found out how passive they had been. For me, this was formative. I vowed that I would be vocal to prevent future wars and address societal evils. Your grandpa and I incorporated a commitment to peace and justice in our marriage vows in 1972.

Most of my life I have been a teacher, writer, and activist. In the 1980s, when the Cold War raged between the US and the Soviet Republics, I feared for human survival and tried to advocate for peace in any way I could think of. I composed songs, wrote books and chancel dramas, gave lectures, participated in public witness for peace, and built ties to national and international organizations that sought ways to end the arms race.

I always believed that addressing the great issues of our time meant joining a movement in which many people participated. I wrote a song based on the German folk tale, "The Weight of a Snowflake," which says that one flake alone cannot break the branch of a tree, but if enough flakes land on the branch, it will eventually fall. We had to contribute to the best of our ability as individuals, but solutions came by working alongside others.

Climate change is about to alter our world drastically. By 2020, many were saying that we had only about 10 years left to turn around a trend in CO2 emissions that would lead to disaster. Already at that time the world was seeing a frighteningly rapid increase in floods, droughts, melting ice caps, extinction of species, deforestation, fires, and devastating storms. I want a world for you that is beautiful, with clean air and water, with biodiversity, with sustainable food production and with just practices that do not harm our Earth.

Your grandpa and I love nature. Part of the reason we moved to California was to enjoy the great sunny outdoors year-round. We cherish vacations we were able to spend hiking in National Parks, snorkeling in the ocean to discover tropical fish, and strolling through Open Space Preserves. The fact that whole species are disappearing and that we as humans have set our planet on a path where it is uninhabitable, with much of its beauty and diversity lost forever, makes me want to despair.

But there is still hope. When I joined the Elders Climate Action network, I offered to provide leadership in Northern California and nationwide by crafting a newsletter that celebrates the small steps people are taking everywhere to combat destructive climate change. I hoped this initiative would energize people, contradict the sense that nothing can be done, and provide examples they could emulate.

If in your life you are given a chance to make the world better, take it. Don't worry about whether you will succeed. Look for others to join you in the effort, and if necessary, plunge in alone until others are ready to help. This Earth is a gift. We are called to preserve it for ourselves and all living things. No effort in the right direction is too small.

With love, Your Oma

Logo of Elders Climate Action NorCal

Protesting Against Fossil Fuel Banks

Part 2

Elders Engaged in Climate Action

Voices:

John Sorensen
Marilyn Price
Bill DeVincenzi
Todd Weber
Clint Wilkins

Elders Climate Action is one of the most wide-reaching climate organizations in Northern California. When I encountered their website online, it felt like coming home to a family that had found its common purpose. In 2022 I joined the Northern California chapter (ECA NorCal) and began contributing to their online climate action reports. Most of the "Interludes" included in this book were initially published as part of my monthly mailing called "Climate Celebrations."

The people I worked with back then continued to be active members. John Sorenson, Lynn Morgan, Marilyn Price, Beth DeVincenzi, and Shirley Freriks are part of the current leadership team. Many others gather with them for monthly online sessions, supporting each other as they share their efforts within ECA and other local, state, and national organizations.

This chapter begins with John Sorenson as the founder of ECA, telling us of how the organization came to be. Next, you will hear voices of ECA members Marilyn Price, Bill DeVincenzi, Todd Weber, and Clint Wilkins.

John Sorensen

What is now Elders Action Network all started on a vision quest in 2011. Life had been good for me, and I wanted to give back in some way. I felt strongly in my heart that I needed to do that.

It took me a while to figure out how to proceed. A key activity was to go on that vision quest in the wilds of Utah. Part of the quest was to spend four days on my own with only water to drink. On my fourth day of fasting, I was sitting on a log when a breeze came by and I heard a voice saying, "Teach leadership." I knew immediately that was the answer.

I hadn't eaten for four days, so I was a little loopy. When I got back to camp and had some food, I started wondering, "Well, what did that really mean?" I concluded that it meant I needed to teach by acting; or, in other words, to demonstrate leadership by leading. This was one of many retreats I went to, and I noticed that other elders just like me were looking for meaning and purpose in this stage of life. It became clear to me that those were the people I wanted to work with.

The following summer I went to another retreat that was led by a shaman, Charles Lawrence, whose mother was from the Black Foot tribe. The next December he was called by members of the Paiute tribe to help them reconsecrate a sacred place called Cave Rock on the shore of Lake Tahoe. Their sacred spot had been overrun and the tribe had successfully launched a lawsuit to get their property back. The place needed to be reconsecrated because it had been polluted with trash over the years. So Charles had come to reclaim the aura of the place and help the Paiutes create a ceremony to reconsecrate it. He happened to have some spare time one day, so living close by, he called me up and asked, "Hey John, would you like to have lunch?" I immediately said yes.

Over lunch, I told Charles about my vision of creating an organization of elders that would grapple with the important issues our country was facing. I said, "But I don't know how to begin. It's a huge goal." He said, "Well, I can help you. Get yourself some friends and we'll form a circle. Then we'll put you in the middle of the circle and we'll help you figure out what to do." And that's what happened. In February I got 12 of us, six men and six women, to meet at a retreat center on the Hudson River in New York. These were people I knew. I chose a special location for this event, an Episcopalian monastery in which I had been for retreats many times before. Some of the people from earlier gatherings came, and some were other friends of mine. When I was told, "Form a circle; then we'll help you," and I said, "Okay," I didn't know exactly what was going to transpire from that. I just went on pure trust that something valuable would happen as a result.

They put me in the circle, and Charles then said, "Now start talking. Tell us what's on your mind." The circle turned out to be a very emotional experience for all of us. And what came out of it was, "Yes, we need to do this, initiate a movement of elders. John, you take the lead in making it happen and we'll support you."

That meeting gave me the sense, "Okay, this is what I'm to do. This is my calling." Those in the circle agreed that there needs to be a movement of elders who have time on their hands and wisdom, who have life experiences and resources to work with, and who have concerns about the future of not only our country but our grandchildren and future generations. They collectively said, "We elders need to be the voice and provide the guidance in creating that movement. So let's get started."

I spent the subsequent year looking for people to get started with this idea. The mission was to form a movement of elders to confront the issues of our time, with special emphasis on the environment and social justice. I went around talking to people about this goal. If their eyes glazed over or they changed the subject, I knew okay, they're not interested, but over the course of the year I managed to get a group together.

We ended up with 47 of us meeting at Mercy Center in Burlingame, California, in April of 2014, to launch the organization. We didn't have a name at that time. We weren't incorporated yet, but the people that were there had heard me. They said, "Yes, I want to be part of getting this started." Nine had been part of the original circle group. There were others from as far as the East Coast.

My wife Sue had helped find and reserve Mercy Center, and had set it up for us. We spent our three-day gathering with the question, "What are we going to do?" Everybody resonated with the idea of elders taking action. At one point two men, Gary Kornell and Paul Severance, came to me and said, "We have a proposal we want to run by you and see what you think." Gary and Paul knew that the second Sunday of September is "Grandparents Day." I wasn't aware of such a designated day, and I don't know if anybody else in the group was either; but they said, "Why don't we go to meet with members of Congress around Grandparents Day? We could urge Congresspeople to pass legislation in favor of the climate—in other words, take some action for the environment."

Gary's daughter worked in the White House, and Paul from Indiana had a background in community organizing. I said, "Well, that sounds like a good idea. It's your proposal. Why don't you run with it?" I knew that giving Paul the lead on that was exactly what he wanted to hear. We did end up going to Washington, D.C., in September of 2015, a year and a half later. We met with 43 different members of Congress. Eighty-five of us went on that trip. We knew this would become a repeated initiative.

Paul subsequently created Elders Climate Action (ECA), a project of our budding organization. In the beginning we called ourselves "Conscious Elders Network." It was the name we voted on, and that's the one that got incorporated. It had two parts, a social justice group and the Elders Climate Action group.

The Elders Climate Action project took off quickly. Grady McGonagill from Massachusetts said, "Let's have a state team in the Boston area." That became our first ECA chapter. Then a woman from Arizona, Hazel Chandler, said, "Well, let's do that in Arizona too." In 2017, we repeated the visit to Washington, D.C., and at that time met with a lot more members of Congress. Twelve of our group were from California, and while there we agreed to create the ECA NorCal chapter. I've been involved with climate action ever since then, but I have to give Paul Severance full credit for getting the project started.

In 2020, we were really worried about the election. About 30 of us met to start an initiative called "Elders Stand for Fair Elections." By November, that project's membership had grown to 700. That's when we said, "From now on, let's call this "Elders for Sound Democracy." So now we had three branches, all under a kind of umbrella organization, now called Elders Action Network (EAN): Elders Climate Action, Elders for a Sound Democracy, and Elders for Social Justice. Social Justice is more of an educational group. It hasn't grown like ECA has, but it's time will come. Democracy is getting a lot of attention these days. With some new leadership, the social justice branch could also flourish.

A wide and determined group of people are now working in the different regional and state chapters. For a while, I led the Northern California Chapter of Elders Climate Action (ECA NorCal), but then I passed on that responsibility to Todd Weber just before the pandemic outbreak. At this time I'm co-leading the Democracy group and participating with the NorCal ECA group. I am also on the EAN board of directors. Apart from that, I'm very active here where I live. We have a group in Truckee called 100% Renewable Truckee. We push our town council, our public utility district, our school board, and our hospital district to advance local climate action. So I'm pretty active. I keep telling myself to slow down, but it hasn't happened yet.

Our country is in an upheaval right now. It's hard to see where this is all headed. It's a big worry for every person committed to climate or social justice. We also still live with the craziness of a potential nuclear war. America's political developments are a worldwide concern, just like we all had a common worry with the COVID pandemic. The whole world's involved in this dangerous phase of history where the US may move to authoritarianism and we risk losing our democracy.

My message to today's youth is this: Stay involved. You need to know what's going on and not turn your back on it. You have to understand that the climate threat is real. Stay informed and think about the future and where we are headed. Build connections because we can all stay hopeful if we work with like-minded people.

It has been a pleasure and a crowning achievement of my life to have participated in creating a movement where thousands are involved in protecting the environment and creating a better world. We are not alone. Now there are other groups, like "Third Act," that bring together young and old to plan for the future. Hundreds of environmental groups are out there. A lot of other groups are concerned about social justice. Our efforts are not in competition. We want to work with each other, and we each have different ideas, but we all share concern for our future, wondering what we are currently doing to ourselves.

We need to wake people up, not in a way that's belligerent or divisive. We can confront each other in a gentle way, saying "You know, take a look at this. I want a healthy world for my grandchildren and for my life. What about you? What do you think?" It's a matter of giving people a voice and then reassuring them by telling how many people are out there, ready to help.

For Further Reflection

- John took time out to go on a vision quest. Do you have ways to put aside everyday responsibilities to reflect on an important new direction in your life?
- Where would you go for such a time of reflection?
- To whom do you turn for support?
- Imagine creating an organization to change the world for the better. What would you focus on?

Marilyn Price

Marilyn is the founder of Trips for Kids ® (TFK), a national nonprofit that has opened the world of mountain bicycling to underserved youth with the aim of building self-esteem, inspiring healthy lifestyles and instilling environmental values. In 2012, she received a Humanitarian Service Award from the University of Michigan for her work. After 28 years, Marilyn retired from the Trips for Kids staff and began working as a volunteer with local environmental organizations. In 2017 she participated in starting the Northern California chapter of Elders Climate Action and has been an active member ever since. Below is her presentation for ECA NorCal from September 2024.

My interest in environmental causes started in 1960 when I was an undergraduate majoring in Sociology at the University of Michigan. We were assigned to read *The Challenge of Man's Future* by Harrison Brown. In his book, Mr. Brown talked about running out of non-renewable resources. I had never thought of this before.

The notion frightened me, and I wondered what would happen to our world. This book sparked a lifelong interest in environmental protection.

A few years after graduating, I became a housewife and mother and was not spurred to action until reading Paul Ehlich's *The Population Bomb* in the mid-70s. I spent the next seven years volunteering for various environmental causes, including for the California Nuclear Safeguard Proposition and for our local recycling center. I developed an environmental class that I ran in my home to teach others about environmental issues and what they could do in their own life to make a difference. From 1982 to 1988 I worked as an administrative assistant for an environmental organization founded by Governor Brown's former Secretary of Resources.

If you are hesitant to take leadership roles because you think you don't have the necessary skills, let me tell you a story. I never had much confidence in myself to become a leader, but in 1988 I started Trips for Kids. I loved biking, had often thought of being a social worker, and cared deeply about our environmental situation. Trips for Kids combined all my interests in one program. I started the program without funds, equipment, resources, or road map to follow. I had to learn everything about running a nonprofit, but I just took it one step at a time. In fact, I never developed a strategic plan.

We were all volunteers for the first ten years. It was run out of my home using my Volvo to transport 11 bikes to the trailheads. One of the goals was to use the bike as a tool to expose youth to the natural world and instill in them a love of the environment and a desire to protect it in their everyday lives. It was not easy. When you put kids on bikes, the thought of learning gets left behind. We had to sneak it in. One time a kid tossed his plastic bottle to the ground. I whipped out my recycling list and let them all know that it could take 450 years for that plastic bottle to go away. I think the message got across.

When we moved the program out of my home, we rented space in a building to open a bike thrift shop. The idea was to reuse materials, thereby reducing landfills and providing affordable transportation to people on limited incomes. It also earned money for our program as donations piled in. Being a saver, by 1995 I had saved enough funds to buy outright the building that housed our bike thrift shop and offices.

We also started an Earn-A-Bike program for youth in an underserved neighborhood near our shop. Just by showing up and learning how to work on bikes and helping customers, they were able to receive job training and earn points to buy their own bikes and parts that had been donated to us. It also allowed us to educate the participants about environmental issues.

In 1999 Trips for Kids became a national program. After 28 years of directing Trips for Kids, I retired in 2015. Trips for Kids had taken 175,000 youth on rides, 68 chapters had been formed (most in the US but a few overseas), and we had 2 million dollars in the bank. So even though I had very little confidence in myself when I began, I learned that staying with my dream and having help from so many along the way enabled me to turn that dream into reality.

I used to tell people that if you can plan a dinner party successfully you can run a nonprofit, and I believe it to this day. But I threw in that to be successful you had to have perseverance, had to work hard, and had to pay attention to detail. And it was always best to be very passionate about your work.

On retiring I decided to devote all my volunteer time to climate organizations. Our environmental situation had become so much worse than it was when I was in college. I never dreamed that it would become so bad, that all life on Earth would be threatened. I had imagined life would go on for millions of years.

Our environmental situation is now a focal point of my life. I have contributed by joining ECA and helping to start our NorCal chapter in 2017. I also started Sustainable Mill Valley, an environmental group that works on issues in my hometown. One of the projects I led was to pass Climate Emergency Resolutions in our Marin Cities. We successfully passed them in 11 of our 12 cities.

I founded the Freedom Singers, a group that brings back the old environmental, peace, and social justice songs from the 50s and 60s that people can sing with us at our current rallies and other events in support of environmental and social justice issues and world peace. In addition, I help many of our local environmental groups. Plastic Free Marin, for example, managed to pass a Reuseable Foodware Ordinance in the County of Marin and all its cities.

The activity that takes up most of my time these days is a Weekly Action Alert that I started in 2017 for our local environmental groups. In July of 2018, ECA started to use it for Action Alerts in Northern California. We call it the "Big List" and are still using it to this day. It contains easy actions people can take, interesting environmental articles, Eco Tips, and a robust calendar of environmental webinars and events. I love working on the "Big List" because it is so important to be aware of the issues and, most importantly, to take action. Many of the action items take only a few seconds to send to our representatives. I have been told that they are tallied, and the numbers are sent to our representatives.

Another way I contribute is by living as lightly as I can on this Earth. In the environmental class I taught in my home in the mid-70s, I shared ideas on how to consume less and why this was important. I had started to live a minimalist lifestyle myself and have kept it up and incorporated more and more minimalist behaviors into my daily life. My daughter and her family share my home, and I now have only two rooms. As I mentioned, I have been a saver all my life and have enough stored in my two rooms to last the rest of my life; so I rarely have to buy anything but food. In 2016 I gave up my car and now get around mostly by bike. I always have Eco Tips in my Action Alerts and have presented some of my lifestyle changes at local environmental meetings.

I have been saving water since our California drought in the early 70s and now work with a group of people to attend our Marin Municipal Water District meetings to urge their Board to place conservation and reuse as a priority in their plans to get more water to our residents. I save all my bath water to flush the toilet and save the kitchen water for outdoor plants.

Personally, I believe in the principles outlined in *Doughnut Economics* by Kate Raworth. It seems reasonable to me that we should progress quickly to live within our planetary boundaries on a finite planet. In addition to no longer using fossil fuels, we need to develop a circular economy, for people in the developed world to consume less, and so much more to ensure that our air, water, and soil are protected and all life on this Earth can continue to exist in this web of life. Doughnut Economics gives us the structures to take actions that are desperately needed.

Our enormous environmental problems are global in scope, and we don't know what the future will bring. I will continue to do my small part for the rest of my life. I don't know if my actions make a difference, but I intend to do what I can with whatever skills I have. I hope that my story will inspire many of you to do more. For those of you who feel you don't have the skills to be a leader, as I did, I hope you give it a try and just take it one step at a time.

For Further Reflection

- Name a book that sparked your interest in environmental protection.
- Which climate organization would you like to join or support in the future, and why?
- Sign up for Marilyn's "Big List" of weekly action alerts. Are you willing to choose one or more options and send a quick letter to your representatives? Links are provided, and it takes only a few seconds to respond.
- Are you considering minimalism as a way to live within our planetary boundaries?

Bill DeVincenzi

Bill was a chief financial officer in the Bay Area for a couple of different companies. Those who work in technology know that a chief financial officer can make or break the company's future. The smartest entrepreneurs in the world can get together, but if there isn't someone who understands the financial aspects of turning what you want to do into reality, the whole gathering becomes a waste of time. Bill is one of those key leaders who can guide organizations to be successful.

Bill moved on to lecturing in the School of Business at San Jose State University. After attending the Al Gore Climate Reality Project training, he became one of the top presenters of the Al Gore climate advocacy content in the San Francisco Bay Area. Ultimately Bill was offered a chance to lead San Jose State's environmental program.

Even average people can achieve great things. The point is to put yourself out there and see what happens.

I've had a long career in corporate financial management followed by another long career in education, teaching corporate finance at San Jose State University. When I retired at age 70, I did some soul-searching as to what I wanted to do with the rest of my life. What came to mind were the climate crisis and sustainability.

I found Elders Action Network at an event in Half Moon Bay. We were given a choice as to which subgroup we wanted to participate in. Climate Change was the one I chose, and so that's how I got involved in Elders Climate Action (ECA), which in turn introduced me to Al Gore's Climate Reality Project.

In August of 2018, I flew down to LA for the training. On the plane, I was having trouble putting my luggage into the overhead compartment, and this nice guy stood up and helped me. It turned out we were sitting next to each other, and I learned that he was going to the training as well. His name was Todd Weber and he and I became good friends. He became a leader for ECA NorCal and has done just a fabulous job. We're all indebted to him for what he's done. One important lesson from that encounter is that you can influence someone just by being kind or by asking a question. Get into a conversation and see what happens.

Since the training, I have done over 50 climate reality presentations. I also started a weekly climate newsletter as I became more educated about climate issues. First, I shared it with family and friends, but now that list of readers is up to about 175 people, including 65 from a different organization.

I decided to found a sustainability group within the community where I live. The Sustainable Villages Community Club was formed, with the intent of making our community as sustainable as possible. After several dozen people had started working diligently on that goal, a miracle happened. Management decided, "Hey, we should have a sustainability director!" So now management took over to make our property and our facility more sustainable. It was a win-win situation. You don't know what might happen when you take the initiative.

As for personal greenhouse gas reduction, I have to give much credit to my wife Beth for what we did. Part of our motivation was to set an example for the Sustainability Club. So, in 2019, we put solar panels on our roof, we bought an electric car, and, you know, off we went. This past year, we added a backup battery. The two things, solar panels and back-up battery, work extremely well together. We power our vehicle and our house that way. We keep an extra battery available for power outages. We installed a heat pump water heater the following year, and in 2024 we bought an induction stovetop. Heating water in thirty seconds is kind of cool! We also have an electric clothes dryer now.

About a year ago we decided to downsize to one car because the other car was sitting in the garage most of the time. Mind you, this decision created a new marriage dynamic for Beth and me, but after 57 years, we figured out that we could manage very nicely with one car.

The last item on our list is our gas heater, scheduled to be replaced in 2025, at which point we will have an all-electric home, which is what our objective was from the get-go. When that's done, we're going to have a turn-off-the-gas party. We'll invite all the people in our club to come and view the facility and join in the fun.

Another area of climate action is influencing policy. Most people shy away from that, but policy is where you can actually gain the most leverage from what you do. You and I can recycle stuff individually, but policy dictates that everybody has to recycle stuff. It has a much greater impact, right?

Some new opportunities presented themselves to me. The first was at San Jose State. A job was posted as director of sustainability. I thought, well, that's kind of what I'm doing here at Sustainability Club. So, I applied for it. I didn't think I had a big chance of getting it because I was 78 years old. Why would they pick me? They usually put young upcoming PhD candidates in roles like that. But, lo and behold, I got the job. I think perhaps my exuberance and what I wanted to do if chosen might have convinced them to give me the job. And it was a fun opportunity because I actually got to do things that one normally wouldn't get to do in a university. I reported directly to the provost, so I had the second-highest person in command of the University behind me.

We created a "Sustainability Across the Curriculum" program where all faculty were encouraged to include sustainability, no matter what their discipline was—engineering, science, whatever. The crowning glory was creating an interdisciplinary sustainability minor that students could choose alongside their major academic field.

The sustainability program also taught me the value of collaborating with nonprofits, corporations, and government agencies. I got them all updated and involved in what we were doing at San Jose State. Collaboration with other groups became a huge part, again, in leveraging influence. The City of San Jose has a climate advisory group that manages the city's clean energy organization. One day, out of the blue, I got a phone call from them saying that I was named Climate Champion of the Year 2022, and I thought, "What?" Here we go again. It was as if the universe was patting me on the back, saying, "Hey, I like what you're doing." These things happen, and you think, "Why me?" I mean, I live in a city of a million people! But that's what happened. What's key is just putting yourself out there, doing what you're passionate about, and really applying yourself, with no strings attached.

When I finished my job at San Jose State, I decided, okay, I'm 80 years old now. I'm retired. This is it. However, never make that kind of a statement because you'll soon find out that you won't be able to live up to it.

In early 2024, the city of San Jose created a climate advisory commission. They put out a note, and I thought, "Hmm, climate advisory commission sounds right up my alley. I'm passionate about it. I've been doing that kind of thing for a while." So, I applied for it. And then a week later, the county of Santa Clara decided to establish a sustainability commission. I thought, oh, that's kind of interesting too. Well, I went ahead and applied for both, figuring that maybe I might get one of them. Okay, as you probably already guessed, I got both. So, I'm now on the City of San Jose Climate Advisory Commission and the Santa Clara County Sustainability Commission. You may ask, "How can you manage to do all this?" Well, remember, I just quit a 20 to 30-hour-a-week job recently. So yeah, I did have the capacity to take those on.

And you know, it's working. I've been on these commissions for a year, and we're making policy, and I think we're doing some good work. So, I'm going to continue to do that. Again, leverage is key here. I can do stuff on my own, but if I do it in unity with others at a level like that, it has a much greater impact. So put yourself out there. If you're passionate about something and really want to get things done, put yourself out there, volunteer and see what happens.

You can donate your skill sets too. When I was introduced to Elders Action Network, I found out they needed a treasurer. I had spent 60 years in that business. And so, I thought, okay, I'm going to donate that skill set to EAN. I've been their treasurer and on the board of directors ever since. But we need a lot of other things as well. We could use marketing skills. We could use human resource skills, a finance backup person, fundraising skills, and board member volunteers.

There is a plaque I keep on my desk. I keep it in front of me because when things get tough or when I get down, this is what I go by. It says, "Nothing in the world can take the place of persistence. Talent will not; nothing is more common than unsuccessful people with talent.

Genius will not; unrewarded genius is almost a proverb. Education will not; the world is full of educated derelicts. Persistence and determination alone are all-powerful." We're all facing a huge challenge with the results of the most recent election. We're going to have to up our game if we want to continue to make progress in democracy and climate. But if we have persistence and determination, we are a powerful force for change. We can be vibrant, inspiring, and impactful.

For Further Reflection

- Do you agree that average people can achieve great things? Give an example.
- Bill emphasizes the need to "put yourself out there." Share a time when you did that.
- How do you initiate helpful conversations about our climate future?
- Have you tried influencing legislative policy? If so, how?
- Bill urges us not to give up when things get tough. Have you ever managed to effect positive change through sheer persistence and determination?

Todd Weber

Todd was my mentor when I first joined ECA. He was available for counsel as I prepared the "Climate Celebrations" newsletter, and he provided me with helpful guidance again as I prepared this book. You will hear his voice in two segments. The first is his eco-biography, where he humbly shares that much of his career was spent in jobs that harmed rather than helped the environment. The second segment delineates the paradigm shift ECA is currently undergoing as we shift away from just dialoguing among elders to tackling the climate crisis by working with multiple generations.

Part 1

The original Earth Day occurred back in 1970. At the time I was a high school student in Waukesha, Wisconsin. Pollution was pretty bad, but our awareness was merely experiential. No one asked what our Earth should be like or whether we should take care of it. We didn't have a sense of the Earth as a living being at that point. We were just goofball graduating high schoolers worried about other things. Earth Day was probably the first event for me that introduced me to the concept of ecology.

For the next seven years, I was in the Air Force, focusing on technology. I did very non-ecological things, like programming for ICBM multiple warhead missiles—all kinds of jobs that human beings should never have gotten around to doing. I taught people how to program the base stations from where the missiles would launch.

After I left the Air Force, I worked in my hometown area in Whitefish Bay, Wisconsin, for a company called Oster, a Division of Sunbeam Corporation, where I was introduced to making consumer products. Next, I found employment in California at Intel Corporation making integrated circuit chips. Next, I moved to Atari Inc., the consumer electronic video games company, where I helped move our consumer game product manufacturing all over the world. Back then companies like Oster, Intel, and Atari didn't pay much if any noticeable attention to the environment.

However, at Atari we made a lot of profit getting kids just to focus on their home TV screens instead of going outside and playing basketball or doing other healthy, important things. Our goal was simply to keep kids indoors sitting in front of their TV screens so they'd send more money to us for more games.

I moved from Atari to Apple, working with Steve Jobs and his team on the early Macintosh computers. There was a growing public sense that big companies are naturally bad and that Apple intended to beat them all with this really cute, smart, little computer. Again, the effort wasn't environmental. It was consumer-focused and commercial. Eventually I ended up at Cisco Systems for the tail end of my career, making sure WebEx (Zoom's principal competitor) was secure and working well. We liked to tell people, "You should use WebEx instead of flying to your next meeting." So we had this little tinge of environmental flavor, but it was basically consumer-oriented, commercial, about making a profit. I spent most of my professional career on commercialization, figuring out how to get people to consume more and more and more.

In 2017, I retired from Cisco, and I had this really crazy thought. As an engineer, throughout my career I was focused on fixing things and on finding solutions to problems. My father was an engineer. I was an engineer. My son is an engineer. My super crazy thought was that as an engineer I could fix the climate. During the last couple of years of my career, I had heard and read about Al Gore, and I had seen his first movie, "An Inconvenient Truth." So I was aware by this time that we had a climate problem. What an engineer does is figure out how to solve a problem and then go, "Okay, I'm done with that. Give me another problem." I had this funny, goofy, stupid idea in my brain that, as a trained engineer, newly retired, I'd work on climate and fix it. That's what engineers do, right? How could anything be wrong with that?

I started gathering information. Unfortunately, at that time climate data wasn't readily available in commercial media. No one wanted to talk about it. After about six months, I got incredibly depressed. I mean really, really emotionally depressed. Here I had retired and was going to help fix something, but it became painfully clear that it wasn't fixable—at least not in the sense that we could just send a bunch of engineers out and they would find the solution and everything would be fine.

Fortunately, at the end of six months of deepening depression, I got a chance to go to the Al Gore Climate Reality Project training. Suddenly, as a good engineer, I was pleased to see a whole lot of data. I got together with a bunch of people who, while acknowledging that climate wasn't going to be fixable in the classic sense, there was all kinds of stuff that had to be done to deal with it. That was a sea change renewal for me both emotionally and intellectually.

As I got on the plane in San Jose to go down to LA for the training, a nice gentleman in front of me was getting himself seated on the airplane. He was a little shorter than I, and he was trying to figure out how to get his hat up into the bin above; so I helped him with that. His name was Bill DeVincenzi. We flew down together, did the Al Gore training together with 1200 other folks, and Bill and I have been friends ever since. Bill began introducing me to other elders working on climate, notably Elders Climate Action - ECA.

I joined the Elders Climate Action group in Northern California (ECA NorCal). During my final job at Cisco, I had managed security programs for different products, so I asked myself how I could best help the monthly NorCal member meetings be increasingly successful. John Sorensen and Shirley McGrath were running the meetings at the time but were also busy with many other activities. John hit a point where he and his wife Sue decided they were going to take a sabbatical. John said, "Todd, while I'm gone, can you keep this running?" One of the main reasons I had to retire from technology was that I could never say no to anybody about anything; I always ended up taking on more projects than I should. So, I said yes.

I had enjoyed managing during my career, and I had been successful at it. Apparently John saw something in my approach that he felt fit with the way that he wanted ECA NorCal to continue to grow and expand. Getting people to engage with each other has always been at the core of the things that I most enjoyed.

However, today as I think about the climate crises we are facing, I don't have much hope. I feel terrible about the commercialization of life, about humanity's view of itself on this planet, and about the ways we contaminate every other society sharing this planet with us.

When I was at Atari and at Apple, I had felt extremely proud of the way that Western society had seemingly turned away from preparing for another world war. It seemed to me like we had done this marvelous thing, converting people into thinking, "No, instead of having wars, just buy the next cool stuff. You'll really enjoy that. You don't need wars. We'll just make some more cool stuff for you." I was thrilled with the idea of turning everybody into a consumer and turning the focus away from wars and from killing and battling each other.

But it turns out that this mess of climate chaos we've created is a direct result of our consumerism, plain and simple, even though yes, the oil companies are also to blame for having known the likely climate chaos outcome all along. When I was in high school and I got my first car, I just loved the idea of being able to buy a cheap tank of gas at the gas station and drive around town all weekend. At most I slightly noticed that burning gas was a problem because when I got too close to the tailpipe, it stank and I didn't like having to breathe that stuff all day long.

Now it has become extremely clear to me, and in retrospect it apparently also was to Al Gore, that we had designed and manufactured a behavior of consumerism. We set up our children from the day they take their first breath to consume, consume, consume throughout their lives. I personally believe that this is what prevents us from being able to deal with our climate crisis. Possibly we never will deal adequately with the climate problems we're creating through our constant drive for more. That's what I hate. I can't change it. I was a part of it. We can't consume our way out of consumerism. It's our entire economic engine across the world today.

Today I see no clear path for humanity out of its destructive loop to destroy the planet we live on together with all other life. We each take our personal initiatives, like putting in heat pump water and home heaters and replacing incandescent light bulbs with LEDs. I try very consciously to minimize consumption. Nevertheless, I still contribute every day to the wrong direction that is driving the world. As a member of this society, pretty much everything I do today contributes to damages tomorrow and beyond, but it already hurts people in the southern hemisphere of our world.

Lately we are experiencing more and more "local" (Northern Hemisphere) destructive severe weather events, and people are saying, "But that's going to raise my home insurance!" I can't easily talk to those people, because what I would say is "Your home insurance is the smallest possible problem compared to those families' homes in Africa (or Indonesia, or Venezuela, or elsewhere) that just floated away in another flood. But hey, you're focused on trying to keep your home insurance low." So that's my biggest personal hope-blocker now. Even though I'm doing as much as I can to reduce my carbon footprint, I can't find words to engage with other people who have not yet come to realize that something is really fundamentally broken. I'm unable to calmly ask, "Shouldn't we talk about that?"

I don't have a message for today's youth. The message about consumption never occurred to me as part of my life-story when I was younger. My father was a German engineer, and he taught me that you always had to have a plan. You always had to know where you were going. But the fact is, you never know where the future will lead. At Cisco, I worked with the guy who created Zoom. At the time, I had no idea that I'd ever get to work on something like that and how it would allow us to see each other today and to have conversations using that technology. Right now there's a guy with six teenagers and young men busily penetrating every single US government technology system they can break into, and the current administration is allowing them to do it. I sure never expected that. Nothing we do now can predict what today's youth will need to do and what choices they'll have to make. And given how badly we elders have screwed up our present world, I'm really glad that we aren't the ones that will make decisions for youth in the future. They'll make something, and it'll be different than anything you or I or even they could imagine today. I'm hoping that future generations will bend the path of humanity away from our current stupidity.

If anything has sustained me these years it's been getting to know the people at ECA NorCal. I continue to thoroughly enjoy how our members talk with each other and share what they are learning and what they are actively involved in.

For Further Reflection

- Are you a person who likes to fix problems? How do you feel when you can't find a good solution? Do you keep trying or do you give up?
- Looking back, Todd admits that some of his professional activities were harmful because they supported our craving for more consumer goods.
- Is there an action you regretted taking in your past life? Did you make corrective changes?

- ❖ Do you think that the path we are currently on is caused by stupidity, greed, misinformation, or apathy—or all of the above? What do you see as the most dangerous aspect?
- ❖ Do you feel pessimistic or optimistic about AI and its potential impact on planetary health?

Part 2

Genetically, I am a tiny bit Cherokee American Indian. I would love the world and future generations to engage with the Cherokee Seven Generations view of reality instead of one that drives us to just consume, consume, consume.

Native Americans lived in tribal communities that successfully practiced sustainability for hundreds of thousands of years. Their communities were constantly engaging at a high community council level, including youth and elders and middle-aged people in their discussions. For example, they talked about how the buffalo moved because they knew that without paying attention and working together as a community, they weren't going to have any dinner next year. Community action was a natural extension of all members' best interests. Seven Generations thinking was therefore part of everyone's daily lives in decision-making and actions.

And so, when the colonists and settlers said, "Why don't you give us your children and we'll educate them?" the Native Americans were right to propose a better deal. They offered to educate the settlers' kids instead. Of course, the settlers didn't think that was a good idea whatsoever. They decided it was better to kill off the majority of Native Americans and take their lands for themselves.

After reading Roman Krznaric's book *The Good Ancestor: A Radical Prescription for Long-Term Thinking*, ECA NorCal's leadership began asking, "How can elders support the next generations when we're a bunch of quite well-off upper-class people? We are the top consumers in the world. We've got nice houses, nice cars, nice you name it. Maybe we need to think differently about being an elder to actually change our behavior and start doing significant things together beyond getting on Zoom sessions and talking with each other about our achievements. How could we make changes for the benefit of youth and future generations?" We can't think forward by looking in the mirror for our behavior and action guide. We fundamentally have to change the way we think.

We scheduled a two-day face-to-face session based on Krznaric's *The Good Ancestor*. At the end of this late-April 2023 session, John Sorensen and Rick Moody were sitting having coffee together, and suddenly it occurred to them, "You know, it's kind of funny. We've just had a room full of elders talking about thinking differently for the sake of future generations, but there wasn't a single youth in the room! Isn't that a problem?" So John and Rick concluded that if we really want to help our elders to think and act differently, we needed to figure out how to integrate with youth for future face-to-face sessions. This launched ECA NorCal into a highly active intergenerational planning and doing process.

We elders tend to focus and look back on the events and achievements of our lives. We often think and speak of our legacies. This isn't really a problem; rather, it is simply a result of our human neural and cognitive development. Roughly, by the time we humans turn seven years old, a structure of how to respond to good and bad events of our young lives has become a neurologically "hardwired" grid in our brains. We humans then spend the rest of our lives filling in the spaces around the grid markers determined when we were seven years old. As elders, we are both blessed and burdened with a rich mosaic of what we did and what we experienced across our lives. Whenever we are asked a question, we typically respond from the perspective of that life experience mosaic, and we can't easily get beyond the neurological limits of our own mosaics.

Since we carry within us these thick and complex elder mosaics, how are we able to help youth create a different life for future generations? Youth are busy actively building their own mosaics, imagining their futures, and experiencing their lives. Youth are not burdened with a whole lot of history to ferret through in order to decide what matters and what doesn't to them. Our youth constantly ask forward-thinking questions, and we need to respect that our experience-based answers may or may not fit or help what they're trying to do. Youth can't spend too much time stewing with us in our memories because they have a whole unknown future of their own to create.

For Further Reflection

- Todd mentions the Native American "Seven Generations" view of reality. What can we learn from their way of making decisions?
- How can we begin making amends for the hurt inflicted on generations of Native Americans?
- As white, well-off people, how do we work on being more inclusive and less racist?
- Todd makes an interesting point about our cognitive development. Given our hard-wired neurological brain grid, established at age seven, how do we best interact with youth?
- We are the top consumers in the world. We've got nice houses, nice cars, and an abundance of creature comforts. As we rethink our contribution to the climate crisis, how can we become true elders and make changes that benefit future generations?"

Clint Wilkins

When Clint Wilkins learned about Future Design, it made amazingly good sense to him. He and youth climate advocate Jayden Wan brought together a powerful team of elders and youth. In October 2024, they organized a retreat and applied Future Design methods to think and talk across multiple generations.

Clint is a perpetual educator. In his long and distinguished career, he served as a teacher, coach, college counselor, dean of students, and high school principal. Since his retirement, Clint has served as an "Encore Fellow" in the Mayor's Office in San Jose, , designing and implementing an intergenerational mentoring program for under-resourced youth ages 14 to 18. On November 14, 2024, Clint gave the presentation below to members of ECA NorCal.

I am a retired high school principal. My main passion has been, and still is, working with young people. In fact, I have spent my entire adult life in their good company. I also have a passion for our planet and the health of our environment. So I'm delighted to describe an event we with Elders Climate Action NorCal organized on October 26, 2024, in Silicon Valley at Community Hall in Cupertino.

Nineteen elders, twenty "mid-life practitioners," and fourteen youth participated. Together we embarked on a day-long journey that's never been done in this country before, called "Future Design." It has grown quite popular recently in Japan but is relatively unknown in the United States. The major purpose is bringing together the generations to explore what our world will look like in 2049. Why 2049? It's exactly 25 years from now, and it also happens to be California's 200th Anniversary.

After introductory remarks from representatives of the three generations involved, we formed groups of seven at six round tables. Each group consisted of two youth, two mid-lifers, and two elders, along with one facilitator who had worked hard in the previous weeks to prepare the event.

Round One. After a warm-up for getting to know each other, we entertained this question: "What is one feature of our life today, in 2024, that would have been unimaginable 25 years ago, in 1999?" The idea behind this was to discover how difficult it is to predict the future. It injected a note of humility from the get-go. As with all the rounds that followed, we used a flip chart to support our brainstorming by taking notes.

In *Round Two* we immersed ourselves in the present using California's November ballot initiative, Proposition 4 (a 10-billion-dollar bond financing program to address climate mitigation and adaptation) as the baseline for what was to follow. Our method was in the form of a debate. At each table, people split into two groups: one for and one against the proposal.

Having prepared our arguments, we launched into a passionate debate. Afterward, we reflected: "What did you learn, especially from the side of the team that does not share your original perspective?"

Next we moved into an exercise we called "time-traveling." We donned ceremonial robes, then formed a human tunnel by raising our arms to form a long arch. Everyone walked through this human tunnel, coming out in the year 2049, gathering around a beautiful globe. At that point we again reflected on the experience.

Round Three consisted of "world-building," immersing ourselves in the year 2049. Wearing our robes from the time-travel exercise, we envisioned our lives leading up to and including 2049. We previewed seventeen futuristic stories leading up to and including

2049, choosing one or two events that were either "cautionary" or "encouraging." Youth, mid-lifers, and elders took turns taking notes. Then we analyzed each of these events through a number of lenses: social, technological, environmental, economic, and political, as well as from the perspective of various actors in these stories.

Round Four. We were now prepared for the climactic moment of the retreat when the actual advocacy for the future would take center stage. In brief, utilizing the funding categories of Prop Four that we had examined in Round Two, three members of each table became advocates for the citizens of the year 2049, negotiating with the other three table members who represented our current year. It became a lively, passionate discussion. Through this negotiation, most of us experienced a shift where the future became the present.

Lastly, we reflected on the action of the day. Each table pulled together what they had experienced and then shared their findings in a plenary session. Before closing, we talked about what we were going to do in the weeks and months ahead. We decided to reconvene in workshop form in early February to develop "future design tools" that practitioners of all ages could employ in their own work as climate advocates. The goal was for them to apply those skills to their actions in local city, county, and state governments as well as in the nonprofit sector.

With our October Future Design Retreat and our subsequent "toolkit building" workshop, Elders Climate Action is moving into new territory as an organization. We're on the verge of scaling up. We've also realized that our true power as elders lies in our role as conveners, bringing together three generations and cracking the code on how elders can be effective changemakers.

For Elders Climate Action, bringing young and old together was a paradigm shift, and working with mid-life practitioners, the ones with the greatest agency, made great sense. They are, after all, what Professor Karthick Ramakrishnan calls "institutional stewards." They are city council members, former mayors, people working with foundations, and sustainability directors.

But we went beyond that age group for our October 26, 2024, retreat. We invited 14 youth interested in environmental advocacy and action. Jayden Wan is the cornerstone, teaming up with another talented high schooler, Audrey Hsu, who is running her own nonprofit called, "Everything Starts Small." These youth activists are remarkable.

As we start the process of scaling up, we are putting to good use the detailed feedback from the participants at our retreat, so critical in refining what we call "the blueprint" and application tools. Over the subsequent six months, we produced a facilitator handbook and a client engagement handbook. These documents position us for reaching out to potential beta testing partners, and we are fortunate to have in place our parent organization, Elders Climate Action, which has chapters all over the country to draw upon. We will continue to refine and improve the materials we already have built.

We're looking forward to working first with other Elders Climate Action chapters nationwide on implementing Future Design methods in their communities. Our power as elders lies in convening the generations to envision a healthier future and as a result taking effective action. We're all eager to take it to the next step, developing skills that will make advocates of all stripes who will be effective in the battle to save our planet from irreparable harm.

For Further Reflection

- Would you like to participate in a Future Design retreat?
- Have you found ways to relate well across generations? What has worked? What are some challenges?
- When you picture living 50 years in the future, what does life look like?
- If you could design such a future, benefiting all life on the planet, can you describe some of the features?

Kids deserve our climate action

Two Short Plays for Street Theater or Group Readings

The plays below, "Carbon Footprint" and "Feel the Heat," can serve as entertaining introductions at a climate event. Alternatively, people may choose to get together with friends and read the script with divided roles.

"Carbon Footprint" raises awareness about our contribution to global heating, but in a light-hearted way. The short scene is fun to read aloud prior to a discussion about the climate crisis. It is intended to invite people to act. It is easy to stage and therefore may work well for a street theater performance.

"Feel the Heat" pokes fun at our addiction to technological gadgets and online games while also raising awareness about the dangers of global heating. It points to our greed and ever-growing desire for "more, more, more." The play lends itself to a follow-up discussion about cutting down on consumption and promoting degrowth.

Carbon Footprint

Two middle-aged persons are sitting at a center table, drinking coffee. One is working on a laptop computer, the other is reading.

ALVA Hey, listen to this. I found this interesting website where they let you test your carbon footprint. *[TRACY ignores ALVA, who tries again, annoyed.]* Hey, listen to me!

TRACY Please don't interrupt me all the time. You know I can't hear you when my ears are plugged.

ALVA It's about your carbon footprint.

TRACY *[impatient]* I am listening to music! *[Takes out ear pods.]* My what? Footprint? *[Looks at the sole of his/her shoe]*

ALVA No, not your real footprint. Carbon footprint. It shows what amount of carbon you put into the air. The things you do that cause global warming.

TRACY I do? Why me?

ALVA Not you personally. We. Most of us. The stuff we should avoid doing.

TRACY *[sighs]* Like what? Give me some examples.

ALVA Well, here. Let's just try and see what they say online. It's kind of a questionnaire to show how we are implicated in climate change. *[ALVA reads.]* They say here that the average carbon footprint for a person in the United States is 16 tons. Globally, it's closer to 4 tons. Sounds like we are the Number One polluters. It's always us in first place, I guess.

TRACY Somehow that doesn't surprise me. Then what do we have to do to get the number down?

ALVA So the goal is ... let's see. *[ALVA checks]* The goal is to avoid a two degrees Celsius rise in global temperatures. For that we need to drop ... *[ALVA looks up]* to under two tons—down from 16 tons—by 2050.

TRACY Sounds like a lot. But at least we have some time then.

ALVA Not really. Right now we are on track to make things worse, not better.

TRACY Surely that's not all up to us? But okay, what do they want to know?

ALVA We have to enter how many miles we travel with our car . . . how often we fly. . . how much water we use . . . how we heat our house. . . what kind of things we eat . . . what our shopping habits are . . .

TRACY Stop right there. I can already see that we are going to flunk this test.

ALVA If you want, we can make some changes in our lifestyle.

TRACY Like what?

ALVA Get an electric vehicle. Or you can ride your bike to work.

TRACY Why me? What about you? You need the exercise more than I do.

ALVA There are also a lot of ideas here on what to do around the house—like turning off the lights, installing low-flow showerheads, turning the heat down ...

TRACY *[exasperated]* Turn the heat down??

ALVA ... use rechargeable batteries, plant trees, line-dry clothing, spend less money...

TRACY *[applauds]* Spend less money! Now there's an idea for you!

ALVA Well, do you want to save the planet or don't you?

TRACY Ok, I agree that we can do better. And I'll work at it with you if you like. But I wonder whether this whole thing with the personal "footprint" just distracts from the fact that fossil fuel companies have been wrecking the Earth for profits, and now they want *us* to take the blame.

ALVA *[looks at her, thinking about it]* Well, you have a point there. We should probably also take a look at what these companies do. And what the politicians are up to. And what the banks are financing. And how other countries are ruining things. And ...

TRACY *[interrupting]* Stop, I get the point. What you are saying is that we should do our part, regardless.

ALVA Yeah, exactly. It seems like it all hangs together, right? Our actions, our choices.

TRACY Let's learn more about what we can do. Maybe we can talk to our friends about it.

ALVA Do you think they'd be interested?

TRACY No idea. But it's easier to figure out what to do if you don't work at it alone. I'm going to try this. *[Picks up the phone and dials.]* Lindy? Hi, it's Tracy. Are you busy next Saturday? I was wondering . . .

Feel the Heat

Two teenagers are poking around on their iPads. The "Chorus" members comment on the action, repeating words that someone shouts.

JESS I found this neat new app.

CAREY Not another one with guns, I hope? I am kind of sick of gunning everyone down.

JESS You are weird. It's just a game!

CAREY It's only a game until we have the next school shooting. Did you watch the news yesterday?

JESS Okay, okay. I get it. But here's another one.

CAREY What's it called?

JESS It's called . . . *[checks]* "Life Points."

CAREY Hmmm. What is it about?

JESS It looks like fun. You can accumulate all sorts of stuff. You know, the latest gadgets you see on YouTube. And the ones my friend talked to me about on Instagram.

CAREY Really? All right, let's try it. At least it's not violent this time, right?

JESS Correct. And it's pretty simple.

[The chorus surrounds them, cheering them on, watching and commenting. The chorus voices overlap.]

JESS You have to press the arrow keys. Get the goods when they come. Quick. Oops! You missed one. Now! There you go! Hey, good deal!

CAREY I have to keep holding down the consumer button, right?

JESS Yes. Over there.

CAREY Okay. Consumer button. Here I go.

JESS Take as much as you can get!

CHORUS Move more quickly. More often. More stuff. *[Voices overlapping]* More quickly. More often. More stuff.

CAREY *[beaming with delight]* Got it!

JESS *[applauds]* Great job! Okay!

CHORUS *[echoes]* Great job! Okay!

CAREY Here's another bonus to collect. Let's play again. *[JESS nods. CAREY pounds on the keyboard.]* There! I won! I won!

CHORUS You won! You won! Go again! Get some more! *[Voices overlapping]*

CAREY I'm getting warm. Do you feel warm? Why is it so warm? And what's this on the screen's upper left?

JESS [*looks at the Ipad]* It's a . . .counter. It counts your life points.

CAREY But look at that! My life points are shrinking. How come my life points are shrinking? I want another chance. Where is my next chance?

JESS There it is, flashing red. Hit it! Grab it!

CAREY *[loud, excited]* I got it! I got it again! *[More quietly]* But ...Why is it so hot?

CHORUS Hot, hot, hot, hot! *[They create a wave, with increasing speed and loudness.]*

JESS Did you see that bargain? It pops up so fast. You should take it while you can.

CAREY I have only a few life points left.

JESS Whow, look at that! Just press it. Go on. Just take what you can get.

CAREY Success! Fantastic. Let's play another round. Great game! *[Beams at Jess.]*

JESS Okay, let's start over and get some competition going. See who of us can get more stuff. *[CAREY sits down at his own Ipad and joins in the game.]*

CAREY I love this. All these opportunities. More stuff, I mean. [*Moves as if itchy.*] I do feel sweaty though. Gosh, it's hot. Why am I breaking into a sweat? How can you stand it? Did someone turn up the heat? Can we turn it down somehow?

JESS *[surprised]* Gee, I feel it too. And now for some reason my escape key is malfunctioning. *[Pounds on the keys]* Stuff just keeps coming. Weird!

CAREY Mine too. I want to stop. *[Increasingly nervous]* I can't stop! It won't let me! Why can't I stop? Now what?

JESS Seems like it's out of control. It continues ... all on its own! The game just continues!

CHORUS *[repeats, making a wave]* The game continues! The game continues!

JESS *[panicked]* GOOD GRIEF, WHAT GAME IS THIS?

CAREY WE HAVE TO STOP IT! RIGHT NOW!

JESS I'M TRYING! I CAN'T!

CAREY PULL THE PLUG! PULL THE PLUG!

Part 3

Leaders Among the Youth

Voices:

Zekun Li
Jayden Wan
Audrey Hsu
Aaryan Doshi
Avroh Shah
Kayla Ling

Getting to listen to youth leaders in the climate movement has been one of the great joys as I put this book together. The youth I interviewed take my breath away with their energy, vision, persistence, and sheer talent. It is amazing to hear about what they have already accomplished.

Several things surprised me. First, I heard repeatedly that we older adults don't need to wallow in guilt. Elders often grieve the fact that the climate and the environment as a whole was destroyed under our watch. We have known about global warming for many decades but continue to ignore it. As members of rich countries in the global north, we greedily enjoyed unlimited access to resources and technology while exploiting the Earth. Now many of us profoundly regret having failed to think about the effect our fossil-fuel consumption will have on future generations.

Interestingly, the young people I talked to don't hold a grudge. They recognize that we face the same reality and make similar choices. Most youth, mid-lifers, and elders engage in activities that undermine the health of our planet. We allow ourselves to be pulled into consumer society. We hide in our comfort zones, ignore the dangers of climate change unless we are personally affected, and let self-interest interfere with global vision.

A second surprise for me was that the youth I interviewed held deep respect for elders and their involvement. They recognized that mid-lifers have limited time to join the movement. They also stressed that today's older generation can contribute much because when they were young, they also faced challenges to overcome. Many were shaped by the Vietnam War, the Civil Rights Movement, and the Cold War with its nuclear arms race. Elders had to learn to fight for our survival and for the protection of civilization. Many became conscientious objectors to war. The future looked bleak even then. And so today, we can bring our stories, understanding, and compassion. Elders and youth can learn from and with each other.

Third, during my interviews with the young contributors, I was impressed by their passion, perceptiveness, and depth of involvement. I admired their technological pizazz and their creative ability to use social media to build a movement or propel an existing one forward. They used various venues provided for them—prizes, festivals, competitions, climate groups to associate with, and even intergenerational exchanges—to build up resilience and community.

My takeaway: The young contributors are fabulous models for the fight ahead. They are not the Pollyanna optimistic type. They research carefully, link up with teams, and pursue roads with steady persistence even though the paths are arduous at times. They bravely take blows and setbacks, but they also proudly recognize achievements—their own and those of others. I still feel apprehensive about the future, but I know that these young leaders will address what lies ahead with creativity and courage. They will be a source of hope and inspiration for people in years to come.

Zekun Li

Zekun Li is the executive director of Silicon Valley Youth Climate Action (SVYCA). He is no longer a youth but a mid-lifer. However, his story belongs into this chapter because he works professionally with and among young leaders as their mentor and co-organizer.

When I was a teenager, I watched the 2004 movie *The Day After Tomorrow*, one of the first films in which they talked about crazy natural disasters happening, like what New York would look like if it froze over. That was a tipping point for me. I wondered whether such weather events could be possible if we're not careful about climate. And when I was maybe 19, I read the book *Six Degrees: Our Future on a Hotter Planet*, by Mark Lynas. It explored the impact of each degree of global warming. It was illuminating and transformative for me.

Silicon Valley Youth Climate Action (SVYCA) was started five years ago by Hoi Poon, a mentor and good friend of mine. She became motivated to engage in climate action after the US withdrew from the Paris Accords. She is a community organizer like me and was galvanized by the youth to do more to protect our planet. So they planned a climate strike with a rally and a march in 2019. Then the question became what to do after participating in direct actions.

I have known Hoi for over fifteen years. When she started SVYCA, she recruited me to be a board member. Community organizing and working with youth have always been passions of mine. I agreed to serve on the board, and a couple of years later, she invited me to become the executive director. It was the perfect time for me. Six months ago, I came on board to help lead the organization to the next level.

Over 100,000 students are linked to the organization through our video work or direct involvement. We now have twelve teams working in 15 cities. It takes time to mentor the students and support them as they grow into leaders for the next generation. All the teams are in the Bay Area, from South San Francisco down to Morgan Hill. We just started a new Fremont chapter last November.

Some of the youth are working in policy advocacy. The biggest effort was to support AB 285, an Assembly bill to put climate into the school curriculum. We had conversations with elected officials to get them to sign on. The bill passed, so now there's a policy to teach about climate change in each school in California.

It takes a whole village to pass legislation, and even more people to get it implemented. California's education system lacks sufficient funding. Even though teaching about climate change is a requirement now, it's difficult to train or retrain teachers. For some schools, the implementation can take a decade. Our Fremont team is currently working on including environmental justice in the schools' ethnic studies program. They are in the process of developing a curriculum that ethnic-study teachers can incorporate. It's not a quick process. They will launch a pilot project this year to try out what can be done.

Our organization is planning our first Youth Climate Leadership Academy for the summer of 2025. For three weeks, five days a week, four hours each day, the youth will focus on climate-related issues. They'll be given carefully curated readings to prepare for class. Every day, students will have small-group discussions to reinforce their learning. We will also bring in professionals in various fields to talk about the current problems and possible solutions. That way the students are not only gaining knowledge about climate but also starting to understand how to address the problems.

Another goal for our class is to teach leadership and advocacy skills. Making things happen in this world is not easy. Politics and money play large roles. However, we will provide training so they can better understand how the system works and what steps are necessary to succeed. It is key to learn skills like power mapping and understanding what levers you can pull to make a difference. There will be formal and informal training at the summer academy.

I personally attend some of our action team meetings. Afterwards I may share suggestions on how we can make an initiative more effective or efficient. Of course, the youth will find their own spin on how to proceed, but often these informal conversations serve to learn about ways to get people involved. Advocacy is most successful when we have relationships. Our civic engagement has to be more than protest or strike or make public statements at a city council meeting. If you want to change the vote of an elected official, you first need to develop a relationship. By the time the vote happens, their decisions have mostly already been made. People will have been involved in backroom deals or have made up their minds before the meeting happens.

Our San Mateo County Student Action Team recently attended a city council meeting where the City Council of San Mateo considered a proposal to remove bike lanes near San Mateo High School. Our students noticed that 90% of the public comments were against removing the bike lanes. Nevertheless, the council voted to remove them. While participating in public comments is a tool to be civically engaged, it doesn't mean that's the most effective tool to achieve your goals.

Community organizing requires you to be proactive in building relationships. That's the direction that we're moving towards as an organization. Beyond just meeting, we can partner with decision makers and influencers on community projects that have long-term impacts. Working on a project together deepens relationships that allow your advocacy to be more effective in the future.

Before my time with SVYCA, our Morgan Hill action team did an amazing job of showcasing the power of consistent advocacy. They went to city council meetings for eight months straight. They showed up for all the meetings to insist that Morgan Hill City Council develop a climate action plan. Their persistence led to the council finally adopting its first-ever action plan. A Climate Action Plan is a document that a city council adopts. It can include many different components, like reach codes, nature-based solutions, switching to more electric city-owned vehicles, whatever it may be. It could be a combination of various initiatives a city can take action on to further their commitment to climate and environmental protection.

So far, the climate movement in the US is far behind compared to other countries. Many people are still stuck in thinking that as long as they recycle, they are doing enough to save the environment. Recycling is not anywhere close to what we need. Europe and Asia are way ahead, not only in transportation or solar installations, but in general awareness. Many people in the US don't believe that climate change is real. There is little in your current education system that promotes climate literacy.

Look at our transit system compared to that of Europe or Asia. In the Bay Area, we are so far behind. It will take two decades to build three BART stations from Fremont to San Jose. Most buses in the South Bay and Peninsula come by every 30 minutes to 1 hour. If US citizens are still struggling with the question of whether climate change exists, can you imagine how far we're behind on any possible solutions?

People don't fully realize that climate change is already significantly impacting our health and that it has to be dealt with now. We can see the impact with the wildfires, or the Santa Cruz pier destroyed in a storm, or San Francisco getting a tornado warning for the first time ever. Those are the things that we can see. What people don't see is the impact of microplastic in our food, water and the air we breathe. Climate change is affecting our whole society now.

Americans are too individualistic in their thinking. I am guilty of that myself. I live in San Francisco, and when a tsunami warning sounded, I thought, "I live on the other side, closer to the Bay. A tsunami is never going to hit me." We need to get over thinking only about ourselves and instead consider the global impact of climate change.

I continue to hope that our education system can get stronger in addressing the issues. What I'm trying to do with the SVYCA organization is to help the next generation of students build a stronger movement. We have 100 active members this year. My hope is we can grow to 200 or 300 students next year. Our main program is with high-school students, but we are also building up our college program and eventually a professional program. After they graduate, the youth will have the leadership and advocacy skills to apply to whatever jobs they choose. We hope they will continue to utilize their skills to organize and bring about changes.

One of our students who graduated three years ago is now at Yale. She's the president of a global consulting program. She's now collaborating with us, sending a group of six students to plan a climate hackathon. At a hackathon, participants are given a problem to solve. It may involve technology, but it doesn't have to. In teams, students can brainstorm. They can involve students and professionals with engineering, design, and product management skills to build prototypes. They may have six hours or a whole day to solve specific climate problems. We need new technology, and if we can partner with people who work in tech, it's an opportunity for our students to make an impression on individuals who may not yet be involved in climate. Perhaps the tech people will start thinking about what the group put together. So we're trying not just to work with our students but also help them understand the power they have to influence other adults. It will set them on a path to build a bigger movement through that process.

Sometimes adults forget just how smart youth can be and how much they can actually do. Unlocking their potential, whether it's middle or high-school students, college students, or young professionals, requires support and mentorship. Passing the baton from one generation to another is crucially important. I hope that more and more people continue to give back and support the next generation by spending time with them to help them grow as leaders.

Learning how to mentor requires tapping into your compassion. You have to understand what students need and help open up new doors for them. Intergenerational efforts take time, patience, commitment, and mutual respect. But the work is rewarding and essential.

The youth voice is largely missing from policy discussions about climate, and youth are underestimated in their ability to unite people. However, when youth share their concerns about climate change and their future, it often helps decision-makers and community leaders reprioritize their agendas to include climate change and increase the budget for solutions.

As students successfully organize and execute projects, they develop the life-long organizing and advocacy skills to make a difference in their community far beyond their four years in high school. They might not be youth leaders forever, but they will be lifelong climate leaders who will influence their community and peers wherever they go.

For Further Reflection

- Summarize the goals for the planned Youth Climate Leadership Academy. What is new in their approach?
- What did students learn when they tried to preserve the bike lanes near their school? How could they have been more proactive?
- Have you been to a country with a transportation system more advanced than ours? How can we catch up?
- Do you like the idea of a climate hackathon? What makes it effective?
- What qualities would you look for in a mentor?

Jayden Wan

I'm a senior at Burlingame High School. Growing up, I didn't think much about climate. I was more into sports and playing video games. But around eighth grade, after the pandemic ended, the pieces came together for me. I was particularly intrigued with politics at the national level and with technology. I learned about social justice causes like homelessness, and I asked myself what we could do about the many problems we see in the news. I figured out that the climate was important and was endangered.

At school I heard about the Youth Climate Ambassador program in the county of San Mateo. I applied and received a $500 grant to participate in a year-long series of meetings and climate actions. The program is still going. It's open to 9th through 12th graders, and I would recommend that everyone apply. It was like a crash course, "Climate Change 101," exploring the different causes, effects, and possible solutions. About 40 people participated during the year. I came away with a greater understanding that climate change is not just global warming or sea level rise but touches on broader issues as well, like other topics I was passionate about, including technology, social justice, and policy making.

Participation in the program included attending slide presentations and lectures, but also interactive activities where we would talk to peers, discuss climate change, and brainstorm solutions. We met both online and in person. The program is based in San Mateo County, but other counties may have similar ones. We received a little stipend, which is a great financial incentive for high schoolers. My brother teased me with the comment, "There are two kinds of green," referring to money and the preservation of the environment.

For me the program was a springing-off point. Right after it ended, I joined another environmental organization that I'm still involved with today. It's called Silicon Valley Youth Climate Action (SVYCA). That one became my cornerstone for getting a sense of what climate advocacy and action are about and how I could participate. Much of the work I've done is at the local level, both for in-person events and online, with a variety of different climate organizations. I have met with people in nearby towns, community centers, city halls, and parks. SVYCA also holds meetings specifically for strategic planning, where it's helpful to see everyone face to face. Sometimes we plan a group bonding event like getting together for lunch or having a picnic.

What I really like about SVYCA is their capacity to achieve real-world impact. Other programs for youth tend to provide only an introductory and elementary understanding of the climate movement. The activities they encourage, like planting trees and picking up trash, are good for the environment but have been around as initiatives for a long time. Until I joined SVYCA, I had never heard of high schoolers engaging with legislators at the local level and working with city councils to pass policy, or hosting school assemblies for climate literacy or clean energy. Those were things that Silicon Valley Youth Climate Action was doing.

Throughout the past couple of years, I've been involved in initiatives ranging from policy advocacy to building electrification. A lot has to do with transitioning from gas to electric, whether in homes, for cars, or for whole cities. I got to know the community better through different initiatives. I was able to host the school assembly at my own high school and brief students there, in collaboration with another organization called Peninsula Clean Energy.

PCE is a community-based electricity program that provides 100% clean energy to San Mateo County. One reason we partnered with them is that they had a vision of applying their resources to the school environment. SVYCA could be the liaison between their resources and our student body. We are trying to make students think about their lifestyles and how they can choose clean energy as they grow up, whether it's for their car, home, or cooktop. As for school staff, faculty, and administrators, we can encourage them to think about their school facilities. Perhaps there are gas heaters that can be replaced by electric ones instead. Those are the kinds of changes that we've been trying to push. The school curriculum is usually defined by the state and then the school district. To my knowledge, not many public schools offer courses specifically on climate.

When talking with youth, we try to make climate conversations as personal as possible. In San Mateo County we don't have massive 100-foot tidal waves. We hear about fires in Los Angeles, but for some people, that seems far away. It's easy for us to detach ourselves from a lot of those climate-related issues just because we don't see them every day. However, we can explain the trends. There will be more and more natural disasters caused by global warming, like hurricanes and wildfires and flooding. So we can ask, "If these are becoming more frequent, could I contribute in any way to try and prevent that? What would be the moral thing to do?"

People can be brought on board with this type of approach. A large majority agree that we should do something to stop our climate from getting worse. The more difficult step is to lead them toward taking action. That's why I try to make the suggestions as personal as possible. I talk about strategies to improve buildings and transportation—anything that involves structures and people and places. People need to engage in climate conversations that intersect with their interests and their concerns.

Say, for example, someone is thinking of becoming a doctor. There is a connection between pollution and public health. Or maybe someone feels that there is too much corruption in the government because of how the fossil fuel industry influences politicians; so they could go into law with a special focus on the environment. High school students think about career options. They like to find out about paid opportunities where they can learn about climate, how it relates to them, and how they can make a difference.

I've been with Silicon Valley Youth Climate Action for about three years now. The organization has made it possible to hone my skills and amplify them tenfold. It's great to be with people my age who have similar passions, similar values, similar drive. It's powerful to be surrounded with good resources and a community of committed people.

I would like there to be more of an intergenerational component. For over a year and a half now, I've been part of an intergenerational future design group that is working on climate advocacy. We held a workshop in February where we invited policymakers, youth, and elders to come together and learn about how they could apply their vision for climate action to their personal life, their workplace, and the different institutions they represent. Some are working in nonprofits, others in education or corporations. We're not sure what will come of our gathering, but it feels groundbreaking. It's something that hasn't been done before.

For progress on climate, technology, and social justice issues like housing, it's essential to have people of all ages working together. There are a lot of venues to solve problems, but you can't just hand them over to young people who may not be in positions yet to implement any of those solutions. It's also a mistake to rely solely on folks between the ages of 30 and 50, the "midlifers" or "institutional stewards" as they are called, because that discounts the valuable perspectives of youth and elders. When making decisions, we must consider how each of these age groups will be affected in the future.

The older participants have a crystallized knowledge based on their lifetime of experience. They have lived through periods that a 17-year-old like myself would only read about in a history textbook. It is valuable to see developments from their perspective. Elders and youth are fairly evenly represented on a global scale. We have to take every generation's perspective into consideration equally. Elders have knowledge, resources, skills, and connections that are vital for the climate movement. Many people in power belong to that age group, and youth may find it difficult immediately to tap into their networks. For advocacy and policy-making, connections and relationships with elected representatives are crucial.

Our polarization today has eliminated many nuances in decision-making processes. Youth need to be encouraged to think about what kind of nation we want to grow up in, what kind of world we want to create. Those are the conversations I want to start: how to grow into good ancestors, what kind of role we want to play in 25 or 50 years. Regrettably, these topics are not commonly talked about among my peers. Part of the problem with long-term thinking is that the world seems to be changing very fast, and it's difficult to get a grasp of what the future holds.

When I think about living 30 years from now, it looks both utopian and dystopian to me. Right now, we live in a time of obvious wrongs—wars, persecution, injustices, and gross inequalities—that seem to get worse day by day. If we know as much as we claim we do, why can't we make changes? The best-case scenario for the future, in my opinion, would be that we overcome tribalism, or "othering," the idea that there's "us" and there's "them." In a better future, we would enjoy diversity in communities in terms of race, religion, and all kinds of other social descriptors. We would show tolerance, acceptance of differences, encouragement of diversity, and respect for each individual regardless of their background or abilities.

In terms of climate or the economy, a lot of our problems stem from the idea that there is an "us" and a "them." Take homelessness and housing, for example. There are people on top and people down low, the rich and the poor, those with influence and those without. A better world would have systems in which all people are represented. The world would be a lot more fair if it were not just based on how much money you have.

Technology will continue to shape the future in ways that can help or hinder attempts to create a better world. A good example is the use of social media. At the end of each class at my school, as soon as people have packed up their things and are waiting for the bell to ring, they whip out their phones and start scrolling, scrolling, and scrolling. The attachment to social media gets in the way of communicating with each other. It eliminates the kinship people relied on in the past, the interactions that build connections and strengthen community. That kind of technology is completely individualized. It confirms our existing biases and prejudices.

Of course, other kinds of technologies have led to vast improvements. We rely on advancements in technology every day. Think of the new heights we have reached through engineering or science data. In terms of climate, though, it is true that technological developments have also contributed to global warming, habitat fragmentation, and the degradation of ecosystems through the advent of machinery and industrialization. You can't fight fire with fire, or in this case, solve these issues of industrialization with more industrialization.

But that doesn't mean we need to give up on making progress. Regarding climate, policy is a powerful tool because it means that we can regulate existing technology to make sure it serves our needs. We can also put limits on where we want certain kinds of technology to progress.

If I consider how I personally want to contribute to restoring and maintaining the health of our planet, what comes to mind is accessibility. I want to make the environmental movement accessible to everyone. And not just in the sense that we put information on a website, and it's there for everyone to see. Rather, we're making it accessible because we want them to join. We want people to feel that they have both a path and an obligation to do so.

Many people are reluctant to vote for climate action because they think it's just a left-leaning idea remote from their daily lives. We need to get people to think personally and long-term about the implications of climate change. One of my goals is to try to convince people of any background that this issue is important and that they can and should take action.

For Further Reflection

- ❖ How do you get people interested in talking about climate?
- ❖ Do you prefer online or in-person interactions? Why?
- ❖ Jayden was particularly drawn to SVYCA because of its initiatives for policy advocacy. If you were to join a climate group, what kinds of actions would most appeal to you?
- ❖ Jayden gives many examples of making climate conversations as personal as possible. How can you do that with your friends and acquaintances?
- ❖ Is your vision of our world 30 years from now utopian or dystopian?
- ❖ What special contribution can your age group make to intergenerational climate discussions and efforts?

Audrey Hsu

Even at a very young age, I enjoyed learning about nature. In elementary school, since first grade, my dream job was to be a zoologist because I loved animals, and so I started reading at my school library about animals and the environment. I remember in second grade coming upon a book about marine biology that had an image of a seal with nylon around its throat. That was the first time I saw something that pointed to our destruction of the environment. The image has stayed with me. I clearly remember that moment when I opened up the book and saw that picture. It just felt so wrong. I wondered, "Why aren't people talking about this more? Do they just ignore it and decide that's somebody else's problem to solve?"

I felt confused, angry, and indignant. I did not know what was going on, and I wanted to understand this issue further. So I started to ask people around me about it. I began reading more about pollution, climate change, and the human impact on the environment. This was still while I was in second grade and was living in Taiwan. As I delved deeper into the issue, I witnessed and started looking into floods after typhoons had devastated rural Taiwanese families, making me all the more aware of humanity's relationship with nature. It was at this moment that I promised myself I was going to be a disturbance to this norm of climate inaction I saw in my community.

As I continued to learn about climate change, I was baffled by how so many schools didn't teach about climate change. The topic simply didn't come up in class. Sometimes we talked in school about "reuse, reduce, recycle," but there was never this emphasis on climate change and its causes as a vital issue.

Because of the lack of climate education, I decided to take matters into my own hands and spread awareness to my classmates. But even my peers weren't listening to me. Some made fun of me. It was frustrating. For years, I kept navigating through environmental information by myself. Throughout middle school I often felt discouraged because it seemed that no one else cared. I felt shut out. There weren't many opportunities for youth to have a platform to express what they wanted to say. I was only 11, and no leadership roles were open for me.

I finally decided that I needed to become a change-maker myself. It could not just end with me finding out about the injustices brought about by climate change and then forgetting about them. In fifth grade I started a club outside of school. I named it "Everything Starts Small" to show that no matter how young you are or how small you feel, you can make a change. I got together with a few others to explore how youth can make a difference. It was a small club, just me and my sister and some of my friends. We launched various projects, like beach cleanups or fundraising through bake sales. Fortunately I had a lot of support from my parents. I would not have been able to embark on this work without my parents' help. For instance, they taught me how to run a meeting and how to handle finances.

Taiwan has indigenous tribes. I wanted to learn more about how they work on sustainability. Before starting our group, I had gone on a trip to a rural part of Taiwan to learn about their way of life—the way they make utensils out of coconut shells and create sustainable soaps and other things. Because I lived in Taipei City, I had to pay for trains to get to the indigenous communities. So I started raising money. For one of the first bake sales, I baked cookies but accidentally used salt instead of sugar. I remember one lady coming back saying, "Oh, your cookies are salty!" And I was like, oh my gosh! Some of our activities were trial and error. But we collected enough money to pay for transport to the rural part of the country.

In addition to learning about indigenous lifestyles, I also went to their communities to teach English. Having grown up speaking English and Mandarin, I could bridge the language gap. Teaching is an initiative I am still involved with, even though I'm in the US now. Some of the people I work with in Taiwan collaborate with indigenous groups, teaching English and helping them use the language to understand some Western culture better, in case they want to move or advocate for changes. The indigenous people are such an underrepresented group in our society. They bear the burdens of climate change but don't have much of a voice.

I moved to the United States going into sixth grade. Currently I live in the Bay Area of California. Other than in the AP Environmental Science class, which I'm taking right now, there hasn't been much emphasis on the climate or the environment. Teachers might talk about it a little bit, but not to a point where students understand the depth of this issue. I still see high schoolers and middle schoolers today treating climate change as a joke when they talk about it. Our education system needs to prioritize the climate topic more because all of us will need to face the consequences of environmental neglect in our lifetime, especially the youth.

By this time, my club Everything Starts Small had grown more and more. In middle school, I decided I wanted to hone in on climate. The group started expanding organically, first by word of mouth, then online. By the end of middle school, we had members in Taiwan, Korea, and the West and East Coasts of the US.

Each week on Saturdays at 6 PM, I run online meetings. It's like an important support group. The club slowly grew into an international organization. My sister is the co-founder, doing a lot of the internal work and updating the websites. I oversee many of our projects. In

2023 we were able to register as an official 501c3 nonprofit with the IRS. It has grown a lot since then.

So essentially in middle school, we began an initiative focused on environmental education. It started with an online blog post focused on empowering student voices. During that time, we had a team in Korea because one of my friends had moved there. The connections spread organically. At the start, I was just looking to see if people around me might be interested in embarking on this journey with me. I wasn't yet allowed to use Instagram or social media. So growing the organization depended mostly on me recruiting and spreading the word directly. I would go to school events to pass out flyers, or I'd pass the flyers out at lunchtime.

This whole journey has taught me not to worry too much about what other people think of me or to care about being judged. Fear of looking stupid or fear of other people's opinions about you really holds you back. And that's why I've learned throughout this experience.

The flyers included some facts about the environment and invitations to join us in our efforts. If my sister had a school event, I would attend even though I wasn't a student there, and I would pass out the flyers and talk to people. Honestly, when I first started doing that, I was pretty scared. I thought, what if they just stare at me, judge me, reject me, and turn me away? But I was determined to stay true to what I wanted to do and what I felt was right. At any opportunity I had, I talked about our organization and our goals. I did that even at an academic summer camp I attended for biology. I approached many people I didn't even know and had never even spoken to, saying "Hey, my name is Audrey; I was wondering if you might be interested in . . . ." Doing that got me out of my shell. Now I feel a lot more comfortable speaking up.

By now we've expanded much more. Once onr nonprofit Everything Starts Small got on social media, it was a lot easier to network. I also have a LinkedIn account that made it possible for many young people to reach out and learn about my work. I realized that other youth also needed some sort of support group or kind of a family. Growing up, I often felt alone, thinking that people didn't care about what I thought; but when I built this group, I was able to create a family for myself and others. My goal is to empower other youth to do the work they want to do in their communities.

We currently have over 150 active members. Everything Starts Small has 23 chapters in 19 countries. We are an amazing team, operating in various areas. For instance, our team in France started because a university student had reached out to me. She wanted to have an impact on the environment, but she didn't really know how to make change. So we worked with her to identify key environmental issues in her area and to grow a team to help her. We empowered her to do what she wanted. In other countries, we are also involved in various environmental projects, depending on what youth want to do. The whole point is to empower the youth, whether through education or other initiatives.

Some members show up for our Saturday discussions. However, since we are such an international group, sometimes it's like 4 a.m. or 3 a.m. for several of our teams because of the different time zones. To deal with that, we use the Discord platform for online communication.

My sister and I also recently built another youth group with about 400 members to support students who work on innovation to make change.

Another initiative was our coffee grounds project. We started in California, but now it's also starting up in France and Dubai. Essentially, it involves collecting coffee grounds from various stores like Starbucks or Verve. Those grounds usually just get thrown away, but they can be used as a slow-release nitrogen fertilizer. We dry the collected coffee grounds and then redistribute it to local gardeners, farmers, and community gardens or to schools with gardening plots. So we're able to upcycle materials in our community to reduce waste, and we also reduce the use of artificial fertilizers and support those who are growing food.

For education, we host programs and workshops in person. We coordinate with various schools, and we teach about how students can develop leadership to make environmental change. We have presented in high school and middle school but also talked in elementary school to younger children. Our goal is to make young people understand what climate change is, why it's such an important issue, why we all need to care about it. All our teams create their our own curriculum. So our team in Taiwan goes to teach schools in their area. Then on Saturdays, we regroup and discuss online. Our California team has created a collection of slides and curricula, including presentations about fast fashion, about how to upcycle materials, how to use innovation and research, and how to use the principles and practices of STEM (science, technology, engineering, and math) to develop solutions to real-world problems.

I am interested in the cross between science and innovation to combat climate change. I feel passionate about using science to create solutions. So in middle school, I learned about biomimicry, which means imitating natural processes to solve problems. And then, in my first year of high school, I did a biometric project. That year, we had a lot of rain in California during the winter. I read about how the immense amount of rain had dumped microplastics into our bay. I also learned about how microplastics affect our health because they are dangerous endocrine disruptors that interfere with our body's system to regulate hormone function. Moreover, microplastics also affect the Earth's carbon cycle. Our oceans sequester a lot of the carbon dioxide we emit, but because of these microplastics, that carbon isn't able to sink all the way down into the ocean. That's a huge issue that isn't talked about nearly as much as it should be.

So during my freshman year, I decided I wanted to do something about it. To start the project, I researched nature systems. I ultimately designed a self-powered microplastics collection device that can attach to the front of boats. I interviewed engineers and scientists to review my design to ensure its feasibility. I also talked to boat owners since they were my target audience. Once I had my research and iterations done, I used my school's 3D printer to build a prototype. Several of my teachers supported the project, and I ended up submitting my work to the youth design competition at the Biomimicry Institute.

I won first place nationally in 2023 for that design. It was an empowering moment. I had dedicated months to that project, and now I knew that I could use my other skills as well. Some people feel that they can only impact the environment if they are passionate solely about climate. Well, that's not true. Climate change affects all of us. Even if you're a writer or are involved in something else, you can use all your other skills to work on the health of the planet. I'm passionate about science, and I'm able to use my science, innovation, and organizing skills to take action on climate.

That's something I've been able to teach many youth through our program. So far, we've been able to teach about 20,000 kids in person through our workshops and through visiting schools. We have also offered online programs closely mentoring about 400 students in their projects. Currently, I still do monthly podcasts and online events with youth to teach about how to use research and innovation.

In my freshman year, I was able to study under a professor from Amherst College, as well as a PhD student from Yale, basically learning about ecology. Then in my sophomore year, I started working in a bioscience lab that is looking into biofuels as a renewable energy source. I investigated how to optimize cyanobacteria growth. Cyanobacteria take in carbon dioxide and then grow in biomass. Using this as a form of energy is way more efficient and better for the environment than conventional biofuels because conventional biofuels are from crops.

I presented my findings at an undergraduate conference last October. Now we're continuing this research. I am also continuing the project with the indigenous peoples to uplift their voices and teach the world about their sustainable relationship with nature. The Taiwan team has visited the villages to learn from them and interview them. We've also worked with some organizations to provide meals in Haiti for those impacted by natural disasters. And we've organized various drives here in California to collect thousands of articles of clothing as well as shoes for our homeless population since they're impacted the most by climate change.

We currently have over 150 active members that I work with in person as the founder and president of this organization. I oversee all the branches and all the projects. Those overseas I support online, but in person I work hand-in-hand with the team. Right now we are raising funds to help the victims of the LA fires. We've assembled a few hundred new hygiene kits and products that we're taking to LA this Sunday.

Climate is an ongoing issue that I feel like everybody should be tackling. It takes hope and energy to carry on. Even when things aren't going well, we must not get discouraged. The climate crisis is not going away. Building my organization has made me feel empowered to continue. My teams have given me a home, a support community of youth my age, to talk about taking action. They feel just as passionate as I do about making change around the world, and they feel supported in their work as well.

For Further Reflection

- Have you observed or heard about instances of animal abuse that horrified you?
- How was the topic of climate change treated when you were in school?
- Audrey relates her experience about not having been listened to. When that happens to you, do you withdraw or keep going? How much are you affected by other people's opinions about you?
- Which persons or groups do you turn to for support?
- Audrey grew into an astonishingly effective activist. Which of her qualities helped her succeed?
- Today's youth build communities for activism via social media. How much do you learn online about climate action? What is your primary source of information?
- Which of Audrey's initiatives did you most enjoy reading about?

Aaryan Doshi

I became aware that our planet was threatened in 2018 when the Camp Fire happened. That was a wake-up call for me. Recently we saw the fires in Palisades and Altadena in and around LA, but the 2018 Camp Fire was much further north, nearer to where I live. In California, we have no snow days when school is canceled; but in 2018, for the first time in my life, schools closed. The sky was just bathed in hues of orange, yellow, and red. The entire San Francisco area looked like pure orange. It felt like you were waking up on doomsday. To see that sort of image just right outside my window made me realize that we had a climate crisis. For me, it was the most evident wake-up call. It made me realize that climate change was really an issue. I think that was the case for a lot of my classmates too. They were like, "You know, wow, what did we wake up to?" I always like to say, if you feel it, then you know it's real. And that was the case in 2018. School was canceled because the air quality was terrible. So it was a momentous occasion that struck home.

Since then, I've wanted to create programs to make people aware of the dangers to the environment. I asked myself, "How can we mitigate the impact of wildfires?" These wildfires ruin the harvest. They ruin crops. I started thinking, "How can we maybe design applications that allow farmers to prepare for natural disasters?"

And so, at the end of middle school, as an 11-year-old, I decided to create an application that allows farmers to be more aware of such situations—a prevention prediction mechanism to help them out. I thought, what if I combine satellite images to create that mechanism and identify hotspot areas where we have soil nutrient deficiencies? That would enable them to optimize crop growth, and we could also use that data to safeguard their crops in the case of natural disasters.

I started by using a platform called Google Earth Engine. It allowed me to use hyperspectral satellite images to receive real-time data on the state of crops. With an index called NDVI (Normalized Difference Vegetation Index), I was able to assign a corresponding number to each crop and see how the plants were affected by such a disaster. That was my project for a science fair competition. It came about as a direct byproduct in reaction to those fires in California.

I am still trying to work out how to distribute the information. How do we reach the farmers and give them the aid they need to combat disasters like wildfires? Is there some sort of indoor space that we could make fireproof? But how would that work, especially if we look at the Southern California region where we have so many farmers who raise our food and provide us with their harvest?

Throughout middle school, primarily because of the Camp Fire disaster, to begin with it was all about wildfire research for me. In terms of communicating the information to the farmers, my first inclination was, you know, maybe I could talk to people at the local firefighter station. So I took some semblance of an app, some semblance of a framework, and I went to ask the firefighters for input because they're the ones who are on the scene, who are feeling the impact in real-time. I sort of pitched the app to them. They thought it was a great idea. I took some photos with them back in seventh or eighth grade. So that was nice.

But then, between eighth and ninth grade, I decided to take a bit broader approach when it came to the environment. And that's how I got into waste management.

My first realization that waste was problematic, and my first attempt to find a solution, came toward the beginning of high school. I was going with my mom to our local computer store and she had this tattered, broken-up old iPhone. A very old iPhone. And I remember the person at the counter telling my Mom, "You know, if you trade in this old, tattered, broken-up iPhone, I can give you credit points for it."

For me as a 12-year-old, that was a pretty preposterous notion. Like, wow, you can trade in this completely worn-out product and still get credit points for it? So I thought, what if we take that same idea and apply it to everyday products, whether that be shoes or e-waste or t-shirts—really anything? That would tremendously help us minimize what ends up in landfills or waste facilities where things are incinerated.

That event was my first exposure to the concept of circularity. What if we apply the same concept by taking a product and distributing it into the hands of a person who may want or need it, rather than having that product end up in a waste facility?

Circularity differs from recycling. We manufacture products with a life cycle of perhaps one to two generations. That means, for instance, after you use a recyclable water bottle, it may end up in a recycling bin, but the chance of it landing in the hands of another consumer is very, very low. Products are from the very beginning not manufactured to be handed down from one state of usage to another without a deterioration in their quality.

In the circular economy, however, we think about how every product can have an infinite lifecycle. That means it's not discarded after I use it or maybe two more people use it. We want to create a chain of cycles where that product never ceases to be used. It's always enclosed within that chain.

I met someone who was even talking about creating a biodegradable pellet. Even after you have plastic in the items you use, it will break down into another sort of substance. On one of my podcasts, someone suggested that this can become the solution to the plastic issues we have today. It is fascinating to see the number of solutions people are coming up with to solve that problem.

That was how I got into the whole idea of circularity, the circular economy. It started with that first incident when I accompanied my mom to the store. From then on, I was like, "How can I sort of broaden this, broaden this reach, create more awareness?"

So I created an organization called Cirfin. For the name, I combined Cir, meaning circular economy, with Fin, meaning financial technology. I added the Fin component because even though today's society is based on a linear economy, we can incentivize the transition from linearity to circularity. That was my primary goal. Creating this kind of organization requires us to incentivize it, because if you don't, people have no natural inclination or motivation to engage in a circular economy. Voluntary participation without incentives is always hard to enforce.

The whole idea behind CirFin is, if we can combine technologies, maybe then we can make a circular economy a reality. For instance, a Blockchain database allows us to keep track of the lifecycle of a product as it's going from person to person. We can keep companies accountable for the greenwashing claims they're making. We can find out whether a certain bottle of tea, for example, is really sustainably made or whether the company is just saying that.

CirFin formed in 2021: Circularity plus financial technologies. Since then, I've been on a journey to reach as many people as I can. I wrote a book about it as well. How can we redesign products with a focus on circularity? It's about outreach, connecting, and painting a vision for a better future.

A lot of people, when they hear "climate," might not understand the issue or have other reasons not to get involved. What I've always tried to do is empathize with the other side as well. For example, often in the media, people advocate for an immediate transition from fossil fuels to clean energy. But it's not as simple as it sounds, right? What about all those who are working in the coal plants? You're just stripping them of a job, and now where are they going to go? That's their livelihood. When I say that I want to reach as many people as I can, it means that yes, I want to persuade them of my message, but I want to do so in a way that takes into consideration where they might be coming from. It's always important to listen to the other side because everybody has a different perspective, a story that you know nothing about. And it's important to incorporate their needs in the mission you're trying to accomplish.

To circulate information, I post on LinkedIn and social media, reach out to friends and neighbors, and talk to just anyone who will listen. I also teach at the local library. About 30 to 40 people come to those events, and I tell them, "Hey, I published this book. Please read it. Tell your friends about it. I brought some copies here, so take one and pass it along." So it started more as a local broadcasting mission, if you will, just in my community, for friends, neighbors, and everybody who showed interest. Then they talk to others in their circle of acquaintances. All the people who came on my podcast were also interested. So whoever I was able to reach with the CirFin community became my main circulators of the book. It's available on Amazon now, called *Designing on the Brink: Products for a Climate in Crisis*.

We are at a pivotal stage now, and building a circular economy is what I feel most passionate about. With every climate intervention technique come ethical concerns we have to consider. Who is it going to be affecting? It's not as simple as an instant snap of the fingers and the solution will just magically manifest. A lot of thought goes into it. For example, scientists have recently been trying to implement a solution called "stratospheric aerosol injection." They propose inserting aerosols into the atmosphere, hoping that we can reflect sunlight back into space and thereby produce a cooling effect that reduces global warming. That's what's supposed to happen in theory. It sounds like a good idea, a plausible solution to bring down our temperatures. But the problem we have to think about is what happens when those aerosols fall back down to the Earth. What happens when they end up in the crops and in our harvest? What happens when it ends up on indigenous land? Those people rely on the land for their livelihood!

So it's always important to keep in mind that our actions may affect people, especially those that we don't see or interact with on a daily basis. They deal with the everyday impacts of the climate crisis in ways that are magnitudes higher than what we experience now. Think about indigenous people who have lost their homes. Imagine your own home, the next day completely gone. That's what they're going through. Think about farmers and others stripped of their jobs. That's what climate change is doing to them.

We need to show empathy and learn where people are coming from, what it is they're doing out in the fields, and how their local environment is affected. Always keep in mind that we currently may not be experiencing the crisis as much as others do. I try to consider the context of the other person and the ethical dimension of everything I do.

There's a famous Indian spiritual saying, "Karmanye vadhikaraste ma phaleshou kadachana." Basically, it means that whatever goes around comes around. It's essentially the idea of karma. That's something I grew up with. In today's consumer-centric world we focus on transactionality, insisting that if we do something, we should get something out of it in return. But we can deviate away from that. Instead of wanting a return, our motivation can be to elevate the quality of people's lives. There should be no numeric quantity or metric attached to it. Those actions are going to be so much more powerful in the future. Years down the line, you realize, wow, I did make an impact on someone's life.

One person I had on my podcast always told me, "You only have to change one person's mind, right? That's enough for you to know that you made a difference. Your impact, in the context of the whole world, may well be small—but if we rally around one another and have each other's backs, we can make tangible, feasible changes." So that's something I always try to stick by: detach yourself from the sense of transactionality and do what you do because you like doing it and you believe in it.

All of us, the young and the old, have a place in this climate action effort. I think the message across all generations should be the same. There is no need to play a blame game, trying to figure out who put us in this state and why. What's happened has happened and you're not able to change the past. I don't think we need to go there. Instead, we need to ask what we can do now. We must be optimistic and think about how to deal with what lies ahead.

Intergenerational climate efforts are good. We need more events where 90-year-olds and 16-year-olds are in a room together to talk about the climate crisis. After all, the youth of today are also participating in unrestrained consumption. Fast fashion is becoming a lot more common among youth. That was something that older generations never even touched on. Compared to younger generations, they lived much more simply. Look at Gen Z right now, the carbon footprint they have, flying in all those airplanes to go from one place to another. So climate damage is definitely inflicted by both sides.

Think about how you can unite. The older people might bring a sense of realism that I think younger generations would value, and the youth could bring ideas. Just intertwine those together. But the message remains the same for all age groups: Don't dwell on the past. Just focus on the future and see what you can do from there.

We have a local Earth Day festival at our library. I'll go there to interact with children in elementary and middle school. Even high schoolers will come out. We'll just bond together and have a good time. For Earth Day, the city of Cupertino will set up an Earth Day festival with lots of booths for information to make our community better. I'm part of a volunteer organization called "Walk Bike Cupertino" which focuses on sustainable transportation in the city. I help set up the specific tent where we talk about walking and biking. At the festival, I'll give information and stickers to people who stop by. The organization Walk Bike Cupertino was founded in 2014, long before I joined. Their whole mission has been to advocate for safety. In 2014, a bike rider from Mona Vista High School was killed when a truck hit him.

The organization was founded because of that accident. We try to minimize fatalities by creating safe biking routes. We want to encourage biking safely to avoid devastating incidents.

We also go to city councils to advocate for creating a more sustainable Cupertino. I've been involved with that for the last two to three years. It's been great to work with people and make that a priority. I always like group settings and face-to-face interaction. For my podcast, we have one-on-one conversations through the lens of a Zoom call since I can't interact with people on a global scale otherwise. But in my home community, I always value in-person interaction a lot more.

Policy stands at the core of successful climate action. The Inflation Reduction Act, the overturning of the Chevron doctrine—those kinds of very important legislative documents are key for initiating and supporting positive climate action. The government can tax these big corporations, encourage citizens to live more sustainably, make sure that we lower our carbon footprint, and invest in a healthier climate future. In the EU we find great examples of circularity. Amsterdam is one of the most bike-friendly cities in the entire world. Someone who I had on my podcast told me that they have bicycling highways where no cars are allowed at all. It's a completely different mode of transportation, but they've created the urban infrastructure to do that. So it's definitely possible, but it takes an initiative from the government to proceed. If we fund inventions like that, we can get to where we need to go. So far, Big Oil companies have prevented progress. The government needs to push forward on a path forward to sustainable energy.

It's not going to be an immediate transition. It is going to be gradual. Jobs will be replaced. So how do we deal with that aspect of it? We must ensure that people still have jobs after a transition occurs. But it's not impossible, right? We've seen it in other countries. After gradual but consistent policy action, they were able to get to a more sustainable environment. In the US right now, policy is the main hindrance to where we need to be.

I have hope for the future though. Let me tell you an analogy I always like to use. I'm sort of an amateur beekeeper in the sense that I volunteer at our local "Honey by the Bay" farm. The reason I first got into it was that bees are one of the best demonstrators of circularity. They show how it works by the way they disperse pollen, connecting distributor to recipient. That's something that no other species on this planet has been able to do. A lot of the circularity we do in society is modeled on them and the way they interact. It's like a feat of teamwork and collaboration. Bees epitomize a circular economy.

One bee in its entire lifetime can produce only one-twelfth of a teaspoon of honey. No more than one-twelfth of a teaspoon! That's just a tiny fraction of what you see oozing out of bottles onto tea and pancakes, a fraction of what you see at the supermarket in rows and rows of honey jars. How did that huge amount get there? It was through a feat of teamwork and collaboration.

I like to think of that in the context of the world. One person's impact is probably comparable to one-twelfth of a teaspoon of honey. It's only if we rally around one another and have each other's backs that we can transform that one-twelfth of a teaspoon of honey into a filled jar. We can make the impact we want and need.

That inspiration from bees has always stayed with me. Whenever I thought I had it hard, a person who I went with to the farm used to tell me, "Just look at the bees and what they achieve each day; imagine the amount of work they have to do to produce even a small amount of honey. It's phenomenal, but they keep at it. They emphasize circularity and accomplish an amazing feat."

I also am inspired by the people who come on my podcast, the way their faces lighten at the thought of creating a more sustainable world. One person from Benin talked about how in West Africa they're transforming plastic water bottles into light bulbs to create solar electricity. There is so much hope and promise in these villages, being able to do something like that.

Just seeing people smile about it, seeing their optimism and enthusiasm, that's what keeps me going. It's the same way that the bees keep me going. The more you surround yourself with people, the less you feel isolated. The tasks feel achievable and the challenges are conquerable. So that's what I try to do: always interact with people and unite as many people as I can. And that's what I think we should all strive to do.

For Further Reflection

- What was the wake-up call that made you aware of the climate crisis?
- What do you admire most about Aaryan's initiatives?
- Are microplastics part of your concern for the future? Have you personally taken steps to reduce the use of plastics?
- What is your reaction to the concept of a "circular economy"?
- Aaryan calls for empathy and paying attention to where people are coming from. Do you know someone who also uses this approach?
- Talk about a time when you changed a person's mind for the better.
- If your community has an Earth Day festival, how can you contribute? If not, can you help start one?
- What would you like to see happen to make your community more sustainable?

Avroh Shah

From a young age, I have enjoyed being in nature: camping in Yosemite with my family, going on hikes and kayaking through lakes. Activities like that have always brought me immense happiness. When I was in elementary school, I started to hear about this thing called "global warming." Knowing nothing about it, I asked my older sister what global warming was. She told me about the greenhouse gas effect and how carbon dioxide is a heat-trapping gas that is overrepresented in our atmosphere as a result of humans burning fossil fuels for energy. She's a very smart person and gave me a pretty good summary of what climate change is and how it affects our lives by causing natural disasters. She explained how it's progressively getting worse as more parts per million of carbon dioxide enter the Earth's atmosphere. That information disturbed me, but I was hopeful that this crisis could be averted given the then recently adopted Paris Climate Agreement.

By the time I had dipped my toes into the world of environmental and climate advocacy and organizing, Greta Thunberg had begun her climate strikes in Europe. At that age, I felt inspired by these older 16-year-olds. I thought the future was in really good hands because we had all these amazing young people speaking up for us. I had also gradually come to realize that as a person cognizant of the climate crisis, I could do something as well. I didn't have to wait to be older to take action and do my part, however small that might be.

So I started a group in my school that was dedicated to conservation and natural preservation locally. For about a year, we organized cleanups in local parks. I also began learning more about climate change and its effects. I was nine years old, reading books larger than anything

I'd read before. I pushed myself to learn these things because I wanted to get a better understanding of something that seemed insane to me. How could we allow our Earth to be impacted like that? I started by reading the picture book version of *An Inconvenient Truth* and other Al Gore publications. I also spent a lot of time on the NASA climate page, looking at different models to see how climate change was affecting our atmosphere and our climate systems.

I decided that my advocacy could best be applied locally through waste management at my school. We clearly were creating too much waste, and I could come up with strategies to fight against that. One idea was to place more compost, trash, and recycling bins around campus. I petitioned for that to happen, and eventually, it did, although it took a long time. I found myself as a nine-year-old explaining to people my age, but also to adults and other figures of authority, that trash in landfills releases methane emissions and that dumping stuff into the environment was no way to take care of the world we live in. The petition I had drafted was easy for students to sign, and I was also able to get several teachers to support me. Then I managed to meet with the school's administrators and persuaded them to install more bins and adopt other waste mitigation strategies. So those were my first big steps into advocacy.

The school I attended at the time used a recycling company that didn't even recycle paper. I learned about a new company called TerraCycle that was just starting up and promised to recycle things that are generally considered unrecyclable. Of course, we now know that recycling is not as effective as it seemed, but at least I had helped implement simple waste management tactics.

In early 2020 I got involved in the Stop Line 3 campaign. I was 11 when I started following an organization called Earth Guardians that I later joined. They were vocal in opposing Enbridge's Line 3 expansion, especially in Minnesota, which is the one spot where the permits had not been approved. I began getting on Zoom calls from around the country in support of this huge campaign that included many Native American water protectors. Earth Guardians was an international group. When a chapter in Minnesota protested the Line 3 expansion on the ground, there was a violent crackdown. I remember hearing about a woman who called 911 saying, "Hey, we're peacefully protesting here, but the police are spraying us with tear gas." And the person responding to her call said to contact the governor if they wanted to stop that police action. What had happened was truly indefensible. Another woman got arrested, and they didn't even allow her to make a phone call, and her friends were unable to contact her from outside the police station.

A different group, called Stop the Money Pipeline, protested against the big banks that were investing in fossil fuels, including Enbridge, the company that was trying to expand Line 3. So I created an informational flyer, printed copies and started distributing the flyers everywhere I could. I went door to door, talking to anyone who would listen. I found that most people had no idea that money from their accounts at Chase Bank or Wells Fargo was being invested in companies like Enbridge. I wanted to raise awareness about our indirect support of Big Oil and to get people to consider changing where they were banking. People weren't expecting an 11-year-old to try to persuade them. Maybe they first figured I wanted to collect donations for the Boy Scouts or something, but here I was, talking about the connection between pipeline expansion halfway across the country and the money they were investing in a bank downtown.

The pandemic was slowing down by then, and my school went back to hybrid learning. I passed out information about the pipeline to my teachers. They were surprised to learn what was happening, which I found to be very funny because I was supposed to be learning from them, not the other way around.

The pipeline campaign ended up in a big loss for us. We were pushing for Chase Bank and a few other big banks not to renew contracts with Enbridge. Without billions in investments from the banks, the Line 3 pipeline expansion would not be possible, just as without investment and support the fossil fuel industry would not be functional. We called the executives of all these banks and wrote letters, but the banks renewed their contracts anyway. It hurts when an effort that you have poured your heart and soul into results in nothing. On the other hand, if no one had protested against the expansion, it would have happened anyway, one way or another. And even though we ended up losing, it was important to talk with people and raise awareness. That created momentum that hopefully may transcend the negative impacts of the pipeline in the long term.

I was working pretty much on my own at that time, although I had joined Earth Guardians and started a small group in Palo Alto with some friends. I did meet a few others locally, who were trying to move the needle on climate locally.

The pipeline campaign technically wasn't over because we figured we could still get either the White House or the governor to reject the permits. A petition garnered around 300,000 signatures. So I hosted an event at my school urging people to sign the petition against the pipeline expansion. I was surprised that the school signed off on it because it was a rather political move, but what was even more surprising to me was that my peers seemed disinterested. The same students had signed my petition for more compost bins, of all things, a few years back. Now they were completely apathetic regarding a much bigger issue. They seemed not to care at all.

I've struggled to understand why so many students my age seem to have just given up. They won't even engage in conversation or do something as small as to sign a petition that opposes perpetuating the harms of the climate crisis. The Greta Thunberg era of climate strikes has essentially come to an end. You still see occasional youth groups protesting for climate, but they don't pull the same numbers that they used to. Back in 2019 more people were in tune, excited, and hopeful. I'm not saying that people don't care about climate change or don't believe in it. I think they have just lost hope.

The problem is that climate change isn't an instantaneous crisis. If it were, people would be out in the streets demanding something be done. However, the fact is that climate change is progressively increasing in intensity. What that means for youth is that we're essentially like a frog in a pot of boiling water. In California, the water still feels pretty comfortable, although even here one can tell that the water is warming up. I mean, we saw the sky turning orange when the Camp Fire burned. We saw massive storms hitting us two years ago, with flooding and trees crushing houses. And yet, people didn't wake up to promote change.

When I speak to my friends, they say, "You're right, climate change is real. It's an issue that I care about. But I don't think I can do anything to stop it." They ask me, "What are you actually accomplishing with the work you do?" Sadly, for the most part, they are right that much of the work I've done has accomplished nothing in terms of actual emissions reductions. So I'm now on my city's Climate Adaptations Committee trying to reduce pollution from transit and housing, which comprises about 95% of our emissions.

Some students are saying, "If climate change is not affecting me directly, I don't really care." They just want to live their lives. I understand that, because even dedicated climate advocates don't often say, "Yes, this is 100% what I want to do in my life." It is true that I have chosen climate advocacy as my path in life because it has become such a pressing problem. But I'm still a kid at heart. I would rather be playing soccer and doing something else. A lot of people my age want to have that freedom short-term. They want to live their childhoods. It's no fun dedicating your life at such a young age to trying to save the world. It may look good on a resume, and people think it's impressive, but It's not a fun way to live because of all the defeats. I've lost ten times more than I've won. The unfortunate reality is the amount of losses severely outweighing any victory we had.

Public support is lacking because climate change doesn't happen instantaneously. If a meteor were hurtling towards Earth, the governments wouldn't be saying, "Let's go make this meteor bigger so it can hurt some more." People would think that's insane. They would be rioting. But climate change isn't a meteor. It's floods, it's natural disasters, and it's science. It seems slow but it's also happening at a fast pace at the same time. The problem is, if we just keep our heads down in the sand, one day it's going to be too late to stop it. And that day is coming very soon, unfortunately.

I have trouble understanding why other people my age are not doing their part to stop the climate crisis. Even the smallest things could make a difference. The truth is, there are very few places that are going to be safe long-term from severe impacts of the climate crisis. In California, we are incredibly vulnerable to rising sea levels. I've seen projections showing that by 2050, parts of East Palo will be underwater. It's happening progressively. Every time we see king tides nowadays, the Baylands flood. That never used to happen.

Last summer, I visited my grandparents in India for the first time in five years. The pollution there makes any wildfire smoke here seem negligible. I went to a national park in India where I saw the mangroves, which are one of the best natural defenses against rising sea levels. I talked with the guide who was steering the boat and asked him, "Has climate change impacted these areas?" And he said, "Yes. These islands that have mangroves are already starting to go underwater. We've already lost islands because of rising sea levels."

People for the most part don't see heat as a dangerous issue yet, but I think the needle is moving on that. They want more tree cover to protect themselves against extreme heat. NOAA ran a heat mapping campaign last summer internationally. A friend of mine created heat maps for urban areas to see who's being affected the most. The maps have shown that poor people and those who live in historically marginalized communities are suffering the most. Instead of tree shade, they have nothing but concrete, which is where heat permeates and then is released throughout the day. People are now seeing the importance of cooling centers to find refuge.

One initiative that gives me hope is the lawsuit Genesis vs. EPA that we filed when I was 14. I was one of the very first plaintiffs to join that lawsuit when I was 13. I met with the attorneys to discuss how climate change is starting to disrupt our lives and is getting increasingly intense. Genesis is suing the EPA to force them to reduce emissions because their discounting practices invalidate the lives of youth.

The lawsuit was brought by a firm called Our Children's Trust. They are dedicated to supporting the human rights of children affected by climate change. They look at the issue not from an environmental standpoint like mine but from a human rights and constitutional perspective. We allege that the government's intentional allowance of deadly pollution levels and climate change is a violation of our 14th Amendment equal protection rights.

We say that climate change disproportionately affects children even though their contribution to it is negligible. Children have no voting rights, so we can't choose an administration that will give us the power to rectify climate change. So we are being discriminated against without having the ability to say no. Therefore we have no other option but to bring our arguments before the court. That's the one podium where children can tell their story and protest. We're suing specifically against the EPA's discriminatory discounting rates and tactics they use to devalue the lives of children and future generations.

They don't see children as income earners, so our lives are valued less than those of adults, using a 2% to 7% discount rate that compounds on itself. When EPA analyzes its cost benefits as they consider different fossil fuel projects, they essentially incentivize short-term benefits over the long-term survival of children in posterity.

The Nobel Peace Prize winner Joseph Stiglitz, who served as Clinton's chief economic advisor, wrote a declaration in support of our case. He said that a discounting rate higher than 0% is discriminatory. It doesn't make economic sense because why the heck would you want to value the life of something that's going to happen in the future as less than something that's happening now or something that's happened in the past.

Discounting originated in the timber industry. When people chopped down a bunch of trees, instead of each tree being valued at, let's say, a dollar, a tree that was older or diseased would be valued as less. So they created certain discounting tactics to price the trees differently based on what they were worth. Essentially, the government does the same, but with the lives of human beings they are sworn to protect equally. So we allege that that's a violation of our constitutional rights.

Sadly, Democratic presidents say that they are for climate action, but when they're in office, they don't do enough. Joe Biden was by far our best president for climate legislation, but the $300 billion he secured for climate was a fraction of the trillions of investments needed. The sum we ought to spend to rectify our country's emissions is going up every day. In 2023, climate change cost the global economy $3 trillion. In

2024, it was $38 trillion. We had projected that sum for 2049, and we hit it 25 years early. A lot of factors contributed to 2024 being an insanely terrible year for the climate.

What gives me hope is that even the rich will one day realize that they can't eat money and that we're not going to be able to escape to Mars. Sooner or later, people will be forced to wake up. We can wait to act until our lives are completely disrupted, or we can act now. That's the choice that every individual will have to make eventually.

People are starting to realize that it's best not to wait to take action. The momentum is building. It might be slower than what we want, but it's happening. The bigger the problem grows, and it will grow bigger, the more we can foster a collaborative mentality that says we can be in this together. It doesn't have to be a dog-eat-dog world. Climate change has the potential to be a unifying factor among humanity.

If I find even just one person who sees things the way that I do and understands the severity of the problem, that gives me hope. There's a potential for a million or a billion other people to join the fight and take climate action. I'm not special. Anyone can do what I've done. Good people are out there, even if not in the numbers I want to see. Any person might one day realize that they can make a difference. Once they do, they won't go back. So at that point, the momentum will only grow stronger.

For Further Reflection

- How would you explain global warming to a child?
- How would you respond if a young person knocked at your door to advocate for environmental protection?

- ❖ What do you see as the role of police during a peaceful demonstration?
- ❖ What actions would you be willing to take to stop the expansion of pipelines?
- ❖ Consider the lawsuit brought by Our Children's Trust. Do you agree that the EPA's discounting rates are discriminatory, devaluing the lives of children and future generations?

Kayla Ling

I am only 16 years old, but my journey into environmental awareness began early. When I was a kid, I loved going to the aquarium, holding starfish and squids in my hand, or picking up worms on rainy days. The natural world fascinated me. My Mom, who is deeply involved in community service learning, was my biggest inspiration. She would organize beach cleanups that made me aware of plastic pollution. She organized activities for the elementary school and even the entire school district. An important part of these events was to combine action and learning. So, we wouldn't just go and pick up the plastic. We would learn from a park ranger about why cigarette butts were the largest contaminant on a public beach.

In middle school, I heard of a local environmental film festival based in San Mateo County, the Burlingame Citizens Environmental Council (CEC) Student Film Fest. The festival invites 4th through 12th graders to enter a film about any environmental topic. I was a 7th grader when I made a five-minute film about factory farming and the pollution from the agricultural industry, focusing especially on the damage caused by cows' emissions. I prepared the content based on Google research, and in the film I was a newscaster reporting about how cows contribute to pollution. I presented my findings with different exhibits, like you would in a legal case, showing how their burping released methane and how their poop affected the local marine environment.

Of course, the cows are not the ones to blame. Obviously, they haven't asked to be raised and bred so that humans can enjoy their burgers. But the way cows are currently used has massive effects on the environment. Our American culture empowers restaurants to create juicy burgers and glamorizes the overconsumption of meat. We are killing forests and chopping down trees that have sequestered carbon dioxide to free up land for growing crops used as cattle feed.

My film, somehow, won first prize at the festival! The Burlingame CEC does a great job of rewarding kids; so I had a little bit of prize money, and the county supervisor gave me a plaque with my name on it. It was a big deal for a 12-year-old. Now, I am an organizer for the festival as my way of saying "thank you" to the organizers for believing in me when I was so young.

The next chapter of my story involves learning about environmental justice. I attend a private school, and I am immensely grateful for our unique learning opportunities. In 7th grade, we had a unit dedicated to environmental justice, a completely new topic for me. I had heard about conservation efforts in the environmental movement, participated in pollution cleanups, and talked about climate change, but I hadn't grasped that social justice and civil rights are inextricably connected to the environmentalism movement.

I was interested in social justice a lot at this time because of George Floyd's murder and the increase in anti-Asian hate. I began to realize that social justice and environmental health are related. Everyone deserves healthy air, land, and water, but our access is not equal.

Minority communities and especially people of color endure far greater climate impacts than whiter, wealthier communities. Looking at the world at large, we see that America, one of the biggest polluters in the world and one of the richest countries in the world, is far less affected than many communities in the global south, Asia, or Latin America. Those other regions suffer from worse pollution and natural disasters, though they contribute little to climate change compared to the US, China, and Russia.

How can we continue to pollute for free when other communities are so vulnerable and already suffering in so many ways? How can we allow that?

I come from a place of enormous privilege. I live in California. I have never had to go to bed worried about food, the roof over my head, or how to pay rent. I always have a warm bed and a safe family to return home to, so the disparities in our world make me angry. The vulnerable communities do not deserve the worst impacts. Injustice is correctable, and we have solutions for it.

In eighth grade we heard from a speaker, climate organizer Daniel Sherrell. I asked him, "What gives you hope when it comes to climate activism? Now that they're melting and endangered, I don't have a personal connection with the glaciers in Antarctica or the pandas in China. I don't feel moved by this." He replied, "Increasing sustainability in our world is not just good for the planet, but it is actually better for the human species as well." That got me interested. He continued, "When we talk about sustainability, we're talking about creating a healthier world with better healthcare, access to green space for everyone, and an education system where kids are learning equitably. Sustainability is a win for the Earth and also for humans." His words excited me and shaped what I wanted to do in the future. They inspired me to help create a more equitable world.

In ninth grade, two big things happened. The first was learning about the National Geographic Slingshot Challenge, an international video competition to which they invited youth from around the world to submit 60-second videos about a climate problem and a climate solution. In their first year, 2023, youth from 80 countries ended up participating.

I loved filmmaking because of my earlier involvement with the local film fest, and I was also curious about the intersections of humans and the environment. So, I created a film about consumer culture: how young people are being pushed today to buy more and more than ever before; how it is glamorized; and how buying hundreds of pieces of clothing online is now seen as enviable. On social media, it is easy to be swayed by influencers who tell you, "You need to buy this. We have a new model of that. Here's the next version of it." What's trendy goes in and out every single day. So you feel compelled to buy the next jacket, shoes, or clothing item. But our consumption comes at a cost.

National Geographic asked for a solution for them to fund. My solution was to create a social media project called No Waste Nation. That effort has been my main focus from 2023 until today. I pour all my energy and vision into it, creating videos and graphics with the goal of fighting waste and overconsumption.

I ended up winning the National Geographic contest. The video they posted on YouTube has had 10,000 views, and views on my account are at over 100,000 across various videos.On social media, anyone can have a voice, and that's where I find a lot of empowerment. So now I have created a video contest of my own. I get to meet people from around the world who submit their videos. The project is not as big as National Geographic's Slingshot Challenge, but I have received entries from Mexico, Perú, and beyond.

Outside of No Waste Nation, I work with a local group called Citizens Climate Lobby (CCL). It is a national organization working on climate action in a nonpartisan way, attending meetings, and lobbying on both sides of the aisle.

For a while, CCL's main emphasis was on carbon fees and dividends. They pushed for a carbon tax because, unlike in other countries, polluters in America can still pollute for zero dollars. The organization has since expanded its policy initiatives, but carbon taxation remains important. In San Mateo County, we have a chapter led by a retired veterinarian, Elaine Salinger. She came into my ninth-grade classroom and said, "I have this local youth group, San Mateo County Citizens' Climate Lobby. Who wants to join?" I didn't know what I was getting involved in, but I raised my hand. I joined as a member and became co-leader of the group a year later.

CCL San Mateo County has a larger chapter and a youth group for 14 to 18-year-olds. I was surprised to find many older people participating. The only adults I normally interact with, aside from my grandma, are middle-aged. But in this climate organization, it is all retired seniors, and they are some of the most dedicated and wise people I know. We are an interesting motley crew of senior citizens and young climate activists. It's a lot of fun. Organizing in community with other people is the best path forward. There's a saying: "If you want to go faster, go alone; but if you want to go further, go together with others." That's certainly what CCL has been for me, as I have learned from other people about what works and what doesn't. We have made mistakes together, but overall accomplished quite a bit as a group. The youth group alone organizes an annual climate rally and has met with Representative Kevin Mullin and Supervisor Jackie Speier to convince them to sponsor bills and support legislation we desire for a more sustainable world.

My path is a little unconventional because using filmmaking and art to talk about climate change is not something that immediately comes to mind as the best way to take action. But it is the best way for me. It is my skill and passion. One common misperception is that the only path forward is going to DC and calling on members of Congress to enact change. We pursue those initiatives through CCL, and they are certainly worthwhile, but I have also found that applying your individual talents to this movement is exactly what we need.

Anyone of any background and experience can join and be of use. If you're a baker, maybe you use sustainable ingredients and plastic-free packaging when you open your bakery. If you're interested in engineering, work on carbon capture and sustainable energy. If you're in finance or work for a company, we need people who introduce sustainable practices and count the carbon emissions to hold each company accountable. If you're interested in filmmaking like me, we need documentarians, and we need journalists to tell the stories of communities. We need doctors to address the impact of pollution on the human body, people to research microplastics, and people to create preemptive healthcare systems that are sustainable and accessible for all.

Climate action will be my focus for the rest of my life. My involvement cannot stop next year when I graduate high school. There's too much to be done. I am counting on younger people to grow up and fill the positions of policymakers—people who aren't just rallying for change but who are the change themselves. The more I learn about sustainability and how wonderful it could be once we achieve it, the more I feel committed. I want to study environmental science in college. Beyond that, I want to continue the activism component, so I hope to continue creating films. As a journalist, I want to write stories about climate change. I envision a career centered around sustainability and climate action because I have found so much community and love while doing this work. I want a better world for myself, selfishly, but also for my younger sisters, my community, and the wider world. I want to speak for vulnerable people who often don't have a voice. If possible, I want the people who inherit this planet to find it better than I found it.

For Further Reflection

- ❖ Consider how parents can encourage even very young children to contribute to a healthy environment.
- ❖ The student film fest organized by the Citizens Environmental Council was foundational for Kayla's development as a youth activist. Do similar opportunities exist in your town? If not, can you help create one?
- ❖ Do you know of a youth who might be interested in the National Geographic Slingshot Challenge? Pass the word along!
- ❖ What do you think of Kayla's No Waste Nation project? What is your own way of holding our consumer society to account?
- ❖ Kayla names film-making as her skill and passion. Talk about another climate activist who contributes as an artist.

Plastic [W]Rap

[Two people dance with robotic movements to reinforce the beat. Clap for each emphasized syllable. The rhythm is fast until it slows in the last third of the rap.]

Plastic all a**round** us,
what can we **do**?
Let's **talk** about pol**lu**tion.
It's **up** to **me** and **you**.

Plastic all a**round** us
is **prac**tical and **fun,**
but **what** are we to **do** with it
when we are **done**?

Put it in the **o**cean.
Put it in the **sea**.
Put it in the **food** we eat.
Put it into **me**.

Clean and **dur**able.
That's fan**tas**tic!
Cheap and **stur**dy.
Buy more **plas**tic!

Throw it in the **land**fills.
Throw it in the **sea**.
Throw it in the **gar**bage.
It's **all** the **same** to **me**.

Household goods un**break**able
made from **plas**tic!
Tableware dis**pos**able
that's fan**tas**tic!

Straws and **food** wraps, **bott**le **caps**
Plastic **bott**les, **gro**cery **bags**,
Cigarette **butts** with **plas**tic **fib**ers,
Furniture and **cloth**ing tags

Plastic. **Plas**tic.
Mountains of **plas**tic.
Maybe not a **good** idea?
Maybe not fan**tas**tic?

Killing **turt**les, **dolph**ins, **fish**,
Oceans **fill**ing with de**bris** —
What about our **own** en**trap**ment,
Plastic **kill**ing **you** and **me**?

Plastic all a**round** us,
what can we **do**?
Con**sid**er the pol**lu**tion.
It's **up** to **me** and **you**.

Plastic, **Plas**tic
No more **plas**tic
Let's **all** come to**geth**er
To **think** what we can **do**.

Global Warming/Warning

Artwork by Shirley Abraham

A message from the artist:

The irony of these works is not lost on me. I purchased the large plastic (fossil fuel!) 30-inch globes online, and all of the plastics are from my own household, in spite of my shopping carefully to choose responsible packaging.

I knitted the yarn scarf using my stash of old yarn, some of which was my grandmother's. I am a grandmother too!

Earth Melting
(Ice Melting on Globe)

Earth Warming
(Knitted Shawl on Globe)

Earth Smothering
(Globe covered in Plastic)

Part 4

Thinking Globally, Acting Locally

Voices:

Carol Vollen
Joyce Mercado
Arlene Kindy
Diane Bailey

Conservation of resources and a love of the outdoors have always been part of my life. I grew up in post-war Germany with parents and grandparents who had survived economic hardship. They perfected the art of making due. My grandparents gardened, canned, and sewed. For me, getting by with very little became a hobby. I was delighted to find ways toward well-being without money. As a child, I never felt poor. My friends and I thrived outdoors, playing with balls, jump ropes, sleds, and skates. I spent many happy hours picking fruit in our garden. In school, no one competed with fancy clothes or accessories. German TV was free of ads, so I grew up without artificial stimulation of what I might want or need. To this day, the push to consume has little effect on me.

In the 1970s I became increasingly aware of the threat to global health through militarism and the nuclear arms race. The possibility of destruction of all life as we know it was real and devastating. I started writing plays, songs, and books to speak up for peace and justice and to promote international understanding.

Awareness of environmental degradation, even without war, came later—both in terms of its danger and my complicity. I checked out a website to test how I contributed to climate damage. On the surface, I looked pretty good. I always walked or rode a bike to my places of employment. My clothing is mostly pre-owned, gathered at rummage sales or thrift shops. I do not throw away food. Peel and other scraps go into a bio container. My diet is meatless. The fruit we eat fell from neighborhood trees or was obtained through bartering. I create my own artwork and make my own jewelry with recycled beads and wire. The furniture in our home consists of pieces that other people no longer want. My husband and I have eliminated flying. We use our car no more than once or twice a week for church, for grocery shopping, or for a hike in a nearby county park. Because of California's mild climate, we avoid heating in winter and cooling in summer. Overall, I try to stay within the limits of my allotment for one of the 8 billion people who inhabit this planet.

However, there is still complicity. I live in a ridiculously affluent society. As a white person, I am privileged. All my life I have had access to food, shelter, people who love me, and excellent education and other resources. Although I don't waste food, I am alienated from its production cycle. I don't know who grew or packaged it and whether workers were appropriately compensated. Some of my food comes in plastic that doesn't recycle and therefore pollutes the land and oceans. As a person living in the richest country on Earth, I cannot deny that my comforts are linked to the exploitation of poor people and resources from the global south. I reside near the super-rich of California's Silicon Valley. My apartment, though humble, came at a price that few can afford. I contribute financially to causes for environmental and social justice, but the suffering of the poor is removed from my daily life experience.

Eco-activism began for me because of a long-growing awareness of living on an endangered planet plagued by selfish human interference with its abundance of resources. I have become convinced that time is running out; that greenhouse gas emissions are accelerating to an uncontrollable level; that all the beauty of creation, which I love deeply, is in horrible jeopardy.

The struggle for environmental justice continues. It is true that we have made progress in some areas. Climate deniers are now a small minority. Thousands of people all over the world have experienced devastating disasters and are working together to find solutions for survival. However, in spite of many steps in the right direction, a stubborn willingness to exploit the Earth's resources continues unabated.

Let's listen to some more voices of people who have taken action.

Carol Vollen

Carol is my upstairs neighbor. There are many environmental groups and organizations with active chapters in the Bay Area, and she is involved in several of them. I admire her kind and calm demeanor. She is an effective communicator and has become a good friend. Aware of group dynamics, she listens well, helps keep the conversation going, and speaks in ways that reflect wisdom rooted in experience. We go to local hunger relief programs together and are part of several activity groups at Menlo Commons, our residential complex. Carol is a wonderful storyteller, often contributing insights or anecdotes that listeners will remember for a long time.

About 50 years ago I became aware of environmental issues. I was living in Los Angeles and I had gotten involved with a group of women who were taking action for positive change. We looked into various issues and for politicians who would support positions that we had taken. And so we were a bit more political than the League of Women Voters—nonpartisan, but political. At some point during that time we decided to have a panel to look at alternative sources of energy. There were seven of us. We considered nuclear, natural gas, hydroelectric, geothermal, solar, wind, and solid waste. Everybody got an assignment. Mine was solid waste, which I thought was incredibly funny. You could make a lot of jokes about that. But it turned out to be really interesting because methane could be captured from solid waste.

Anyhow, we had this panel where we presented our findings on several different occasions. So that was the beginning of my journey. I didn't know that much about the threat of climate change at the time. Now fast forward to 2016. I heard about this group that was initially called the Conscious Elders Network. It's now called Elders Action Network. I think they had started two years earlier. There were two primary groups: Elders for Climate Action (ECA) and Elders for Social Justice (ESJ). I became involved with ECA.

We went to Washington the following April around Earth Day. We held a conference with Ralph Nader as our featured speaker, and we lobbied Congress. Or rather, we attempted to. We went to every House and Senate office, but we mostly just left literature because there was nobody to speak with. There was actually some kind of a vote going on and everybody was in the chambers for this vote. And then on Saturday, we marched. There were, according to the Washington Post, about a quarter of a million of us marching in downtown Washington, D.C.

That was where I first heard about Al Gore's Climate Reality Project, which became my next thing. In October of 2017, I went to Pittsburgh for the training. Al Gore began doing these trainings all across the nation and even all over the world and did not charge participants for attending. You had to pay your way there, of course, and you had to find your accommodations. All they wanted in return was a pledge that you would take the climate message and spread it. We were there for three days, hearing from amazing speakers. There was so much information. And Al Gore was there; he was very much a part of it.

By now, many of the members of Elders Climate Action have done the training. I came away with really valuable information. Afterward, I began to make presentations. This was still in early 2018. Today, in California, there's no longer much doubt about climate change. People are certainly more educated about it. But back then it was important to awaken people to the dangers ahead. I spoke in churches and synagogues and other groups that were open to learning more about the climate crisis.

I had been living in Burlingame, California, for decades. While I was working full-time, I didn't pay that much attention to Burlingame politics. However, after I stopped working, I became active in environmental advocacy. I called a woman that I knew who was very involved in the politics of Burlingame and asked her if anything was going on environmentally. She said, "Oh yes, we have a Citizens Environmental Council. Come and see." So I did, and I became a part of the Council until recently when I moved to Menlo Park.

As a Council member, I found out about what was happening not only in Burlingame but in the whole Bay Area. Menlo Park was a shining example. I mean, they were advanced beyond most of the other cities on the Peninsula. I'm talking about San Mateo County in particular.

I also took steps to focus on climate in my personal life. Still in Burlingame, I gave my home a complete environmental overhaul, making it more climate-friendly. I installed drip irrigation and put drought-resistant plants in my garden. No lawns, of course. I got an electric car and I put a charging device in my garage. I wound up eventually switching to LED lighting everywhere. I bought an induction stove and installed a heat pump water heater and a heat pump HVAC system. I put in solar panels. I got rid of gas utilities altogether and totally electrified my house.

And then I would tell people about that. At this point, I was just being an individual contributor, basically wanting to reduce my carbon footprint by going electric. When I talked to people about what I did, they were usually receptive. Whether they were interested in changing their lifestyle, that I can't speak to. I know people who made changes but others who didn't.

The future is very scary. I mean, consider the fires in the Los Angeles area. There'll be more fires and floods in the future. And heat. Global warming is going to be terrible. I don't know what it will take to convince people.

At least these disasters are making it impossible for people to ignore the problem and dismiss the link to climate change.

Recently I attended a group discussion about climate, and I said I feel like I need to apologize to God (or whatever created this incredibly beautiful Earth home) for what we as humans have done to this planet. We should be taking care of our Earth. Instead, we're screwing it up. I feel such sorrow about that.

And I am worried about my grandchildren. I mean, I have grandchildren now who say they don't want to have children because they're so concerned about what's going to happen. I don't even know how to respond. I just listen. I am usually an optimist, but when it comes to where our climate is headed, I struggle. We clearly are not going to meet the goals that have been set. I'm not going to be around to see the worst of it. That makes the pain even greater in some ways.

However, there are some good things. Youth vote in very small numbers, but those who are registered vote in higher numbers in favor of preserving the environment. Youth are the ones that have to drive climate activism. Not that I'm passing it off to them or saying that we don't still have an obligation. We certainly do, but they are fresh voices.

Also, in California, we have pretty aware leaders. Our politicians tend to care about the environment. And a lot of organizations are oriented toward the planet's preservation. I don't know whether there'd be a way for them to come together to coalesce and become huge. It would be nice if that could happen. In California, efforts are underway all over to enact laws and encourage environment-friendly practices. Everybody I know personally is aware and concerned. I just don't know about all the other people, even in this state.

I'm originally from Chicago. One time I was back in Chicago for a wedding and my older son was with me. Chicago had just had a horrendous heat wave and over

100 people died, mostly older people alone in apartments without air conditioning. At the wedding, my son was sitting at a table with younger people who said, "How can you stand the Earthquakes in California?" And he said, "I've never been hurt in an Earthquake—but you just had a heat wave that 100 people didn't survive. I've never been in an Earthquake in California where 100 people died." You know, people weren't at that point making any connection between heat waves and climate change.

A couple of years ago, I was in New Orleans for my son's birthday. It happened to coincide with Mardi Gras, so we had a lot of fun. But I couldn't help but be horrified by all the garbage, all the trash that got dumped on the ground—you know, the beads, paper, junk—the amount of trash that collected in the streets was horrifying.

Let me give you another example of trashing the planet. My daughter is an environmentalist. She's also very much a water person, an ocean person. When she goes to the ocean, she finds crap in the water. I remember one time—I was probably snorkeling—she did repeated deep dives to collect garbage. She wasn't using scuba equipment. She could hold her breath long enough to go way down, pick up trash, and bring it back up.

Plastic bags are another nightmare. Of course I carry non-plastic bags in my car and I take them into the market when I shop for food. Plastic pollution is a huge thing. And it's really hard to avoid using plastics. Eventually it ends up in the ocean and it ends up killing creatures. I saw a picture just a few days ago of a penguin with some kind of plastic wrapped around its neck. Turtles get trapped too.

Here is a more positive story from when I was still walking around my area in Burlingame a lot. I had had a hip replacement; so I had one of these devices that let you pick up things without having to bend down. I was past my hip replacement problems, but I used the device to pick up garbage. Somebody who knew my daughter saw me and said to her, "I think your mother was walking with a cane." So my daughter called me, and I had to reassure her that I was just collecting trash.

Future generations are going to look back on us and think, "How could they have done all this damage?" If I could time travel and speak to people a hundred years from now, I would apologize to them, the same way I would to God. I understand their anger. I feel the anger myself. Unless something drastic is done to change the trajectory, the Earth is going to be unlivable. The heat alone will make it unlivable for so many people.

In closing, I just want to say that I feel a lot of sorrow.

For Further Reflection

- Carol gives many examples of how information can change into action. Trace how she got started with various initiatives. Then share the beginning of your own journey as a climate advocate.
- Carol mentions plastics and trash as her "climate nightmare." What is yours?
- Do you feel that we need to apologize for our past inaction and environmental abuses? If so, to whom do you direct that apology, and what does it sound like?
- As you consider the climate future, what scares you most?
- As you consider the climate future, what makes you hopeful?

Joyce Mercado

Joyce has lived in Alameda for over 30 years and has been recognized by the city for her climate work. Her book, Save the Planet in Your Spare Time: A Climate Protection Handbook for the Busy Person, was released in 2024. She has also developed a toolkit for ordinary people to adapt to climate action in their home communities. The following talk was presented at an ECA NorCal online event on January 10, 2025.

I am always excited about presenting to people who get things done. I hope to inspire you with some fresh ideas for measures you can take to protect the climate.

The 2023 Annual Report of the Intergovernmental Panel for Climate Change motivated me to write a book. It was a particularly dire report, indicating that to hold the temperature increase at 1.5 degrees Celsius and thereby avert the worst of climate change impacts, we would have to reduce our emissions by half by 2030 and get them to zero by 2050.

We don't have much time to meet these goals. So I thought to myself, think globally and act globally instead of just acting locally. Maybe there's something I can do to have a wider impact. Why not write a book about what I've been doing in Alameda? Perhaps that will give others ideas for initiatives they can try in their towns to protect the climate. I had to give myself a little pep talk before starting because I'd never written a book before and didn't know how to do it. I told myself, "You can do hard things, Joyce. Other people have done this. Just dive in and try." So I did. I learned as I went along and got the book published in May of 2024.

But what got me passionate about climate protection in the first place was reading a series of articles about climate change in the *National Geographic* magazine 21 years ago now. They included this very nice chart showing the correlation between the concentration of carbon dioxide in the atmosphere in blue and the temperature of the Earth in white, over hundreds of thousands of years, with current-day carbon dioxide concentrations off the chart. That made me realize that we had a very serious urgent problem. The destructive effects of climate change were not way down in the future like I'd been imagining.

So I decided to start by systematically looking at my carbon footprint and making changes in my life. The very next day I pulled the bike out of the garage and started riding it for errands around town. At first, it was just to go to lunch or gather with friends. But then I put baskets on it to go grocery shopping or to mail packages at the post office. Next I put lights on it so I could bike at night, which I found was very relaxing because there's less traffic and it's quiet and peaceful.

Having eliminated unnecessary car trips, I then replaced all the CFL light bulbs with LEDs. I got a home energy audit from our local power company, and they suggested additional things I could do with the house. They also pointed out other improvements I could make.

Then I thought, when I do have to drive, I want to drive efficiently. Air pressure in the tires can decrease gas mileage up to 3%. You can save quite a bit by ensuring your tires are pumped properly. Anyone who has ridden a bicycle with low pressure in the tires knows how much more energy it takes to drive like that. The same applies to cars. Also, I cleaned out a bunch of excess stuff I had stuffed in the trunk of the car because the more you haul around, the more energy it takes to drive the vehicle. And I carpooled for kids' activities, like for my son's little league and soccer practices and my daughter's gymnastics practices. Carpooling cut way down on the need for personal trips. It saved a lot of time too because parents would help each other out with going places. Another interesting side benefit was that you could hear much more about what was going on in your kids' lives because they talked to each other in the car as if you were not there. You could be more in touch with what your kids are doing, especially during their teenage years when they don't communicate very much.

One last thing I learned was the UPS trick. When you set up a route to make several stops, go in a clockwise fashion so you're making right-hand turns instead of left-hand ones. That way you're not waiting for oncoming traffic to clear, and you don't have to wait for the light when turning right. That's what UPS has implemented in their vehicle routes, and we can do the same thing.

Between bicycling, walking for errands, and making driving more efficient, I went from filling up the tank every couple of weeks to every six weeks. So it's quite effective.

I also wanted to focus on reducing, reusing, recycling, and composting. Everything we purchase takes energy and therefore contributes to emissions—to get the raw materials, manufacture the item, ship it, and ultimately dispose of it. So the less stuff we use the better.

I was already doing very well with recycling and composting, but I could up my game on "reduce and reuse." So I got a reusable water bottle, a reusable coffee mug, cloth napkins to replace paper napkins, and dishcloths to replace paper towels. I haven't bought those paper products for decades now.

A friend told me that the "reduce, reuse, recycle, and compost" advocates have now added "repurpose" to the list of actions that will help lower the amount of stuff going into the dump. In California we are passing laws on the "Right to Repair"; so the mantra becomes "Reduce, Reuse, Repair, Repurpose, Recycle and Compost."

As for junk mail, I got rid of all the catalogs that came to our house by calling up the companies and asking them to remove my name from their list. Catalogs are an ongoing battle because every time you buy something new, they add you to their catalog list. The only solution is to keep fighting. The amount of junk mail we receive now is way down.

Another thing I did was switch to a primarily vegetarian-based, vegan-based diet. I started by eliminating beef and lamb because of the methane emissions cows and sheep burp up and because their manure releases powerful greenhouse gas emissions. Animal agriculture also leads to deforestation as land is cleared for grazing and growing animal feed. At first I replaced beef and lamb with chicken, turkey, and fish, then slowly migrated away from those animal products and finally eliminated dairy by switching to almond milk.

I put up a clothesline. I hung laundry before I went to work, and when the kids home from school, it was their job to take it off of the line. They whined a little bit, but it didn't kill them. This arrangement saved a lot of our gas-dryer usage. In addition, I switched to using only cold water in the laundry because most of the energy for doing a load of laundry goes to heating the water. Today's detergents are formulated to work just fine with cold water. It's more gentle on the fabrics, so the clothes will last longer as well.

We replaced our gas dryer with an electric dryer trying to electrify as much as we could at home. We installed solar panels on our roof to conserve energy. I made the most recent change a year ago when I bought an electric vehicle.

Having taken all those individual initiatives, I thought, "Well, I'm just one person. How can I influence others besides leading by example?" First I wrote down all the changes and made a checklist. On one side of the page, I wrote down climate-friendly actions. On the back, I listed resources to help you do the activities. When I got together with friends, neighbors, and colleagues, they often wanted to do more to protect the climate, but they weren't aware of the many things they could choose. So that's why I created this checklist. I started handing it out to everybody I knew. I would give it to parents at soccer or gymnastics practices. They were appreciative, except for one person who refused to take it because she didn't believe in climate change. Everyone else loved it. I've given out thousands of these now. People are grateful, so that's encouraging to me. I also posted the link online to invite others to download it and customize it for their city. That way people could make it relevant for their particular communities and embrace different changes as individuals or groups.

All along, as I became a climate activist, I changed my habits and added more initiatives. I would write letters to the editor of the local paper about the actions I was taking and the benefits I was receiving. To my surprise, they published each one of my letters. I carefully followed the submission rules. (My book includes suggestions about how to write a good letter to the editor.) I wrote so many letters to the editor that one day the editor called me up and asked if I wanted my own column in the newspaper about climate change! That's how I became a columnist. It's great.

Writing a letter to the editor takes only about an hour of your time. You reach thousands of people because subscribers love reading letters to the editor. So you're affecting a wide swath of people with very little personal investment. Moreover, politicians also read the letters to the editor to keep a pulse on what is important to their constituents. That means you are also influencing your local city council members, state representatives, and senators.

After I retired and had more time, I went to the Al Gore Reality Project training where I learned how to give effective presentations on climate change. At my subsequent talks, I covered basic facts about rising temperatures, adding a little bit about the science of climate change, the global impacts, and some local solutions. And then I would close by giving out my checklist so that people had something constructive they could do.

I've given a couple dozen of these presentations. One day when I was visiting a winery, the winemaker asked me what I was doing in my retirement. When I told her about my climate talks, she invited me to present in her winery.

Going through your daily life, if you see something that needs to be changed, you can be the one who takes the first step toward changing it. For example, I noticed at my son's Little League practice that there were no recycling or compost bins. They only had garbage bins. I thought, well, maybe I can do something about that. So I looked up the head of the Little League contact information online, called him, and told him what I was interested in doing, and that I wanted to help them. He said, "Well, you'll have to talk to our garbage guy, Tony." So I called up Tony. It turns out that he was not particularly interested in recycling and composting, but he served as a volunteer because he wanted to keep the kids safe from vermin by keeping the fields clean. He said, "If you help me, we will implement your project."

He needed labels for the different bins to show people what to put where. He did not know how to sort. He wanted a list of the waste items that you get from the snack bar, with instructions on what goes into which bin. So I sent him an email with that information, and I made the signs to glue to the bins. He ordered the bins from our local garbage company, and away we went with the implementation. This whole initiative took me only two phone calls, one email, and making some signs. It was not that big of a deal.

That's how it often is. You think, "Oh, that's just too much to take on," but it may not be very much work at all. It was very rewarding to see parents and kids go up to the bins with their items, read the signs, and figure out where to put what. That way we were educating them about what to do at home as well. And the really cool thing is that Tony got super into this recycling process also. He and his son would police the bins, and if someone had put something in the wrong bin, they would fix it. When he went to the snack bar, they convinced him not to use Styrofoam cups. He grew from someone who knew nothing about recycling and composting to someone stopping waste at its source.

Here is another project I took on. The checklist I had prepared mentioned many ideas about resource conservation and saving money. I reached out to the food bank to see if they would be willing to hand out the checklist to clients when they come to pick up food. I also approached the local power company to see if they'd be willing to give each client a compact fluorescent light bulb. They agreed, and we ended up handing out a thousand light bulbs and checklists to all the food bank clients. Again, this just took me printing up some checklists, taking them to the food bank, making a couple of phone calls, and seeing that the project was implemented. It didn't require all that much effort.

Partnering with other people is a good thing. I organized tree plantings with my fellow Rotary Club members. I consulted the Alameda Recreation and Parks Department to determine in which park we could plant trees, and they offered to install the irrigation for the trees. And then I worked with 100K Trees for Humanity to source the trees for us as well as teach the volunteers how to plant the trees properly. Lastly, we picked a date for the Rotarians to show up and do the planting together. We do this once or twice a year now. It's an effective way to get more trees in the ground. And the Rotarians love it.

Our Rotary Club now gives away tree saplings at our Earth Day Festival each year. The first time we did that, we used oak tree saplings. That effort had limited success because not everyone has room for oak trees. So we learned from this experience, and the next year we gave away fruit tree saplings instead. Those went like hotcakes. Everyone wanted a fruit tree. The end effect was that we got a lot more trees planted in Alameda in the yards of homeowners.

So far I've been talking about individual behavior and how to modify it. But policy changes matter as well. It's important to support candidates who protect the climate. The minimum is voting. In addition, with a little bit more time, we can pitch in by helping them with their campaigns, walking a precinct, texting, writing postcards, and donating money to their campaign. All those I've done. And then, once we get them elected, we have to hold their feet to the fire to get climate change policies implemented.

There is an app called Climate Action Now. Once you fill in your zip code, it knows your representatives and gives you all sorts of standard pre-written emails to send about upcoming bills and about your passion for addressing climate change. That's a very effective tool. I use it every morning for about five minutes to shoot off 10 or 15 emails.

My main message is to be bold and act for the climate. Just because you haven't worked on that so far doesn't mean you can't do it. If you see something that needs to be done, you can be the one who does it. Don't hesitate, and don't procrastinate, because thinking about it tends to take longer than just doing it. When I wrote my book, I spent more time procrastinating than actually writing it. You too can save the planet in your spare time. Be bold.

For Further Reflection

- Do you know people who "get things done"? What makes them effective?
- Share an example of thinking globally and acting locally.

- Which of Joyce's initiatives have you not yet taken but would like to try?
- Joyce's interaction with Tony illustrates how one little step or friendly conversation can lead to positive changes in the local community. Do you know a similar story?
- Joyce's message is, "Be bold and act for climate." What is yours?

Arlene Kindy

My husband Cliff and I decided to live below the federal taxable income level because such a high percentage of our income taxes pays for war. We understood that this decision would have consequences. It meant we wouldn't be able to do some things because we didn't have money for them. However, it freed us up for other things we could do. We had more time to volunteer, more time to maybe meditate instead of working nine-to-five jobs. We got involved in disaster work, rebuilding after hurricanes or other crises. In addition, we volunteered in the church, taking on various assignments. Cliff joined Christian Peacemaker Teams. So we had opportunities and time to do other things besides just work, because we didn't need that much income.

Then we decided to plant an organic garden to raise our food and take produce to market. We did that for many years. It allowed us to be at home and to have our children with us. We didn't have to pay for child care. Our two daughters were with us when we worked at home or did volunteer service.

The Church of the Brethren has a disaster ministry headquartered in New Windsor, Maryland. They send out teams, mostly after the "muck and gut," as we call it, is done. We clean up and rebuild. My job has mostly been cooking for the teams that go out, and Cliff goes as a construction leader. So we've been trained to serve in Brethren disaster ministries. We go for a month, sometimes two or three months at a time, to help the teams that come. We get a different team every week, from various places across the nation, mostly through the Church of the Brethren denomination, although other people are also welcome to participate, of course. The program provides food and housing, and the volunteers work for free.

Cliff and I have been doing this for 50 years, in lots of different states. We've been to Michigan, Florida, North Carolina, and elsewhere. We worked in Kentucky after floods and tornadoes. We rebuilt a church after it burned down in South Carolina. The program sent us to Samoa one time after a tsunami and to various places after hurricanes. Most of the assignments were domestic. Once we went to Guatemala years ago, where we joined Salvation Army volunteers. We worked there, the same as we do here, building houses after an Earthquake.

So yes, Earthquakes, floods, tornadoes—that's where we go. After the national news comes out, people and the disaster relief ministries hear about what's happening. Then they contact the worker teams—Mennonites, Brethren, Samaritan's Purse—to respond with whatever is needed. Usually Cliff and I go later, not right at the "muck and gut" time.

I see a link between the actions of Christian Peacemaker Teams and climate change. If we can stop wars and oppression, which is what they focus on, then we can have less junk in the air, as it were. I see it from the angle of consumerism. I mean, wars consume resources. People fight and die. Making weapons takes a lot of energy and resources.

We try to live simply and consume less. When we moved to Indiana, we bought a farm together with another family. We decided to grow our own food because that way we know what's in it. And we wanted to do it organically, not using chemicals that are often made with petroleum products. Opposing wars, living simply, and taking care of the environment are all connected. Everything is connected. So then we used compost piles and sequestered carbon to help the environment.

When the war started in Iraq, Cliff decided to stop tilling the garden with a gasoline-powered motor. So then we just used our backs and tilled with hoes. That was a little step that we took to keep the petroleum out of the air. On our farm we also have grass patches that we don't mow because it's not necessary to mow everything all the time. Now we have an electric solar-powered mower to avoid using petroleum energy.

For years we sold some of our produce at the local farmer's market. Some veggies were kind of shaped in a way that we figured people wouldn't buy; so we gave them away or sold them at a reduced price. People saw it as a little gesture of goodwill, but it also helped the environment. We need to learn more broadly what's good to eat and what isn't. It will take a different worldview. If a fruit or veggie is imperfectly shaped or has a spot on it, isn't that okay?

This all has to do with consumerism. We try not to consume as much, or to buy, buy, buy. Limiting waste, for example, involved not just giving food away, but in general not living wastefully. Personal choices make an impact on the community. I have a little saying on the fridge that says, "Use it up, wear it out. Make it do or do without." That guideline applies to anything—clothing, food, household items. Learn to live with a little inconvenience. It will work just fine.

That also applies to your living area and daily activities. We live in a refurbished drive-thru corn crib. Cliff can talk about that more. We have a "solar dryer" for clothes, which would be a clothesline. We drive a plug-in car. It's a used hybrid car, so it takes some gas, but around here, we can go to town with just electric power. We have a solar water heater in the summer, so we don't have to use petroleum to heat our water. In the winter, a wood stove with copper coils on the side heats our house and the water too. We shop for clothes at the thrift store, wear them out, and then use them for rags.

Also, I have convictions about washing clothes. We have a ringer washer that uses only one tub of water to wash the whole load of laundry. You start with the whites and then, as you get to the muddy clothes, you use the grey water for the dark clothes. We rinse in well water that you don't have to change for each load. We want to be clean, but it's not necessary to shower several times a day. We dry off with a clean towel, but the towel doesn't need to be changed every time after we take a shower.

Those are just examples of using less and less—not being so fussy about being super clean. We don't need to be germophobic. We can use other people's cast-offs, be they clothes, food, or tools. Some people change the color of their kitchen decor or throw things away routinely. I don't care what color my dish towels are. For those kinds of things, we're willing to use other people's cast-offs because that way they don't end up in a landfill.

When we go to see our daughter in New York, we take the train most of the time. It's a better mode of transportation, less wasteful than a car that holds only the two of us. For entertainment, we mostly just watch the birds instead of paying for other entertainment. We spend money on bird seed and enjoy watching the birds when they come to feed.

In the garden, we mulch by putting straw or old hay down between the rows. That holds the moisture in, so we don't need to water. We have also put out huge milk tanks at the corners of the barn to collect rainwater. When we do need water for the garden, we can draw it out of there rather than the groundwater.

We're surrounded by fields using four center-pivot irrigation systems that draw from the aquifer. Any time there is a little bit of dry weather, the farmers turn the irrigation on. Thousands and thousands of gallons of water are dumped on the mono-crop fields.

During periods of drought, our garden doesn't suffer much because we have made the soil rich. It has a lot of compost and humus in it, so it holds the water. We take care of the garden before the drought comes, and then we are ready with plenty of water when we need it. Of course, we don't use fertilizers. Petroleum is usually used to make fertilizers. We only use compost and cover cropping, meaning crops that feed the soil. We often use soybeans and rye for cover cropping.

I think a crucial way for people to preserve the environment is to lower consumption. I wish people could learn to deal with a little inconvenience—not in the sense of carrying our water from the stream, I don't mean that. I mean, for example, that you don't have to go shopping for only one thing just because you want it at that moment. We live five miles from town. If we drove to town and back every time we wanted some item, that would not be very environmentally correct or helpful. It's like, okay, I can just wait a little bit. I don't have to have everything right now.

We need, first of all, not to consume so much and just be more patient with our wants. Take time to think, do I really need this item? Also, what we buy doesn't have to look like Saks Fifth Avenue. It's okay to be out of fashion. One doesn't always have to wear fancy name-brand clothing, you know. For me, what others think about how I look and what I wear isn't as important as learning to have meaningful relationships.

My choices don't just apply to my family or my house. They affect the world over—the people in Vietnam, Cambodia, South Africa, and other places. When you buy clothes, remember that a lot of clothes are made in sweatshops around the world. So we could just recycle the ones that are already around instead of always buying, buying, buying, new, new, new. And when I use petroleum products a lot, it messes with the air, and the air is for everybody, not just Wabash County.

I try not to buy at Walmart because they have taken over the mom-and-pop shops where people used to get what they needed. And Walmart buys goods really cheap from around the world where people have a lot less than we do. When Walmart started, they said, "Oh, it's mostly American-made," but it wasn't. When we exploit others in the world for our convenience or our consumerism, I don't think that's healthy for the planet. These days we go to wars over what we want to have or consume, and we abuse the Earth for the same reason.

My message for young and old is "Learn to live with less. It's okay to be inconvenienced every now and then. It's okay." That also applies to eating lower on the food chain, not needing to have a steak or pork chops every week. Eat more fish and vegetables because all that meat production in factory farms pollutes the air and the water systems. And we don't need that much protein. A lot of people in the world live on much less. We can eat beans and rice, and that's good for us anyway.

Here is what gives me hope. We just got an email from a woman who said, "I've been hearing about you. I want to come and learn more about sustainability from you." We get those requests fairly often. People want to visit the farm and see how we do things. It doesn't mean they want to do it exactly the same way, but it might spark an idea in them that they can apply when they get home. I find hope in that, the fact that people want to know how to live more gently on the Earth.

For Further Reflection

- Consider the steps Arlene and Cliff took to reduce their carbon footprint. Which of their actions are too radical for you? Which could you imitate?
- How can we best help after a disaster?
- Explain how wars cause environmental damage.
- Talk about your attempts to lower consumption. What did you learn? What was the most difficult?
- If you want to learn "to live more gently on the Earth," whom do you turn to for guidance?

Diane Bailey

As a sustainability programs administrator at Palo Alto Utilities, Diane Bailey helps run some of the home electrification programs for the city. Earlier in her career, she worked on environmental justice in the San Francisco office of the Natural Resources Defense Council, trying to curb diesel pollution through statewide regulations and legislation. Then she led Menlo Spark, a local organization with the mission to help the city of Menlo Park become zero carbon. The goal was to put policies in place to make the city zero carbon by 2025 and to become zero carbon in reality by 2030. Most of the policies have been adopted and approved by the city council. During her eight years with Menlo Spark, the organization put groundbreaking measures in place, including local zoning ordinances with renewable energy requirements for new buildings in a new green district.

Sometimes I feel like a grain of sand on the beach of the environmental movement. It takes thousands of people working together to effect the change we need. You have to view your work in that context. One person can't stop destructive climate change, but if you have tens of thousands of people working together toward the same goal, then maybe we can.

I got into environmental work in college. I learned that in my hometown of Binghamton, New York, several water wells had been designated as a Superfund site. The Environmental Protection Agency had found that thousands of sites in the US had been contaminated by hazardous waste being dumped, left out in the open, or otherwise improperly managed. I read in a news article that two of the four wells providing the community's water had become a Superfund site because the water was polluted.

When I was a teenager, we moved to a new home that was served by the contaminated water wells, and we were told not to drink the water; but we all showered in it and used it. Teens are not always obedient. Later I began to have chronic nosebleeds and headaches.

When I went off to college, my nosebleeds and headaches stopped. Perhaps it was a coincidence; who can say. But for me it was shocking that even though the community leaders in the government knew that hundreds of families were being poisoned by the water, they didn't do a lot to protect people. There were whispers about not drinking the water, but everyone ignored those and kept using the water. I couldn't believe that such contamination and inaction to prevent people from being poisoned could happen in this country.

Then I learned that all across the US, hundreds of communities had been affected. Remember *Civil Action*, the movie with John Travolta? That's one example. Another is *Erin Brockovich*, showing how industrial use had poisoned water. No one intended the disastrous impact on health, but it happened.

In college, I wondered what was being done to protect those communities. That started a lifetime of environmental activism for me. In graduate school, I focused on techniques to clean up water. I went to graduate school in Houston, Texas, which is the largest petrochemical refining center in the US. The biggest issue there is air pollution and how it affects people with chronic illnesses. On bad air days in Houston at the time, children were streaming into emergency rooms, unable to breathe.

That opened my eyes to the whole field of toxic chemicals, not just through water, but through air. I realized the damage the use of fossil fuels created for the surrounding communities. So I dedicated my life's work to reducing the pollution created by using fossil fuels. That's been the throughline of my career.

My degrees are in chemical and environmental engineering. My first job after grad school was with a law firm in Houston, Texas, where I served as their science or technical advisor. I'm not an attorney and I've never practiced law, but I've worked with attorneys for many years, helping them provide the technical justifications for cases they were working on. One of my first cases involved suing chemical companies for damaging the people's health in the city of Port Arthur, Texas. People in Port Arthur who lived near major chemical companies and refineries suffered terribly from health impacts because of the air pollution. Those companies settled the case, which was a big win for that community.

To prove our point, we pulled demographics from the US Census. Nothing was online at the time. It turned out that the community was 99% African American. The demographics showed that this was a classic case of environmental injustice. People facing hazards were minority folks living near the plant, not the corporate leaders who lived comfortably far away and were largely white. Minority populations were paying the price. We documented the health impacts of those facilities on low-income families. We held gatherings and worked with church leaders and members of their congregation to learn how people were affected. It was sad to hear how many families thought that rashes were normal and that respiratory illness was just part of growing up. Getting people to sign on to the suit was tricky because Port Arthur was a company town, meaning a lot of the folks were employed at those businesses that were harming their health.

The work in Texas set me on a course to focus on environmental justice, especially on cleaning up toxic air pollution. In 2001, we moved to the San Francisco Bay Area where I worked at the Natural Resources Defense Council.

While in graduate school I saw some outrageous injustice in the different wards (political and geographic districts) of Houston. Climate change, however, wasn't treated as much of an issue. There were guest lecturers, of course; people talked about the topic; but in the late 90s, it was still thought of as a faraway phenomenon. We figured, "If we reduce our energy use just a little, if we're a little more efficient, we'll be fine." The climate wasn't thought to impact humans. We connected the issue more with those polar bears up north we saw in heartbreaking scenes on TV.

But gradually the climate came into focus for me professionally. The reason is that the intersection between air toxins and climate change is large. The same source, namely the use of fossil fuels, is driving both air pollution and climate change. Our health is at risk right now because of air pollution, but this pollution will also damage our climate in the future.

And here we are in 2025, and that future is now. I never thought in grad school that I would witness the impacts of climate change firsthand. It has been painful to realize that you worked your whole professional life towards solutions and did not get far enough. We are not making good choices as a society. Many of the impacts were preventable, but we chose not to act in time, and now we're paying the price. You see that with the terrible fires in Los Angeles and other disasters.

A small handful of companies have created such outsized damage, not just to our climate, but to our health. Every day we burn their products, and the combustion byproducts cause air pollution that harms people. Statistics show that globally millions of people die every year from air pollution caused by burning fossil fuels. Studies identify "global disease burdens" based on international collaboration among scientists. The documents show what they call morbidity and mortality—sickness and death from different causes. Air pollution is one of the biggest drivers of global morbidity and mortality. Most people don't realize that. I think that heat is the biggest driver in terms of total deaths from climate-related catastrophes. Fires, floods, and extreme weather may play a bigger role economically, but in terms of public health, it's air pollution and heat. The "global disease burdens" study puts mortality from air pollution at around seven to ten million per year. Not all of those deaths are caused by fossil fuel, but most are. It's millions of people every year.

I spent about 25 years in advocacy doing everything I could. There comes a point when you realize maybe it's time for fresh ideas. Maybe we need to pass the leadership on to the next generation. I wish the youth didn't have to deal with these issues, but the fact is that a lot of the strategies we tried didn't work. Maybe we need new folks to take control and try other strategies.

Our country did not get major US climate legislation until 2022. There was a climate bill in Congress earlier, but it failed, and that was devastating. I worked for a national group that had been promoting carbon cap and trade for a long time. Had we started implementing that 30 or 40 years ago, in the 90s, it might have been successful. Now it's kind of a bad joke. I wish we had made a more concerted effort to get off of fossil fuels, including natural gas—and maybe calling it methane or fossil gas rather than natural gas, since that's what it is. Maybe then we would be less dependent on it today. But the fossil fuel industries have had their thumb on the scale of our political system for a long time, and so many of the attempts to legislate our way out of the crisis by moving to more renewable energy and getting off of fossil fuels have failed or been too slow.

What sustains me is the activism of the youngest generation, the teenagers of today, leaders like Greta Thunberg. There are so many young climate activists. They are finding creative new ways to advocate for climate via social media. Watching how the young generation gets active gives me hope.

For Further Reflection

Diane would never make the statements below. Explain why each one is false or dangerous.

- The Earth's resources are there for us to use as we see fit.
- The price we pay for something reflects the full cost of making it.
- Everyone can make it if they just work hard enough.
- Racism is only personal, not institutionalized.
- We have "throwaway" resources, species, and people.
- There is some place called "away."

A Song About Making a Difference

The Weight of a Snowflake

"Please tell me the weight of a snowflake, my friend,"
a sparrow asked a wild dove.
"Oh, I know it is virtually nothing at all,"
said the bird on the tree branch above.

The sparrow said, "Well, let me share an event
I observed some time ago.
I sat near the trunk of a fir tree one day
when it suddenly started to snow.

It was not a blizzard, it was not a storm—
more like a dance it would seem,
without any violence, but steady and calm,
falling lightly, as if in a dream.

I counted the snowflakes on needles and twigs,
having nothing better to do,
and soon the big branch I was looking at
carried three million seventy-two.

When the very next snowflake dropped on the branch
of the tree so proud and tall,
though it weighed almost nothing by itself,
I saw the branch break off and fall."

Having shared her story, the sparrow flew off,
and the dove said, "I no longer doubt
that perhaps it will take only one person's voice
for change to come about."

Part 5

Building Community

Voices:

Rachael Londer
Folake Phillips
Jordi Espin Marti
Elaine Salinger
Teri Whitehair

We live in a time of great uncertainty. Over the last few decades, technology has revolutionized how we think, spend our time, and communicate with each other. Each innovation has opened new doors, but as we embraced it with enthusiasm, we also found out that it came at a price (literally and metaphorically). Many people are spending life in isolation, glued to TV channels that provide endless entertainment and a barrage of misinformation. There is diminished interest in getting out of the house. The COVID pandemic intensified our withdrawal from community life. Online chats and calls replaced face-to-face encounters. The convenience of staying at home led people away from direct human contact. These days, many can no longer imagine living without constant access to social media. How essential is it for humans to socialize with each other directly rather than through a screen?

When John Sorenson started the organization Elders for Sound Democracy, a team formed called Civic Values and Culture. It appeared that we had lost our civility towards each other. People could no longer find common ground, even with former friends and family. How could we construct a healthy society in this context? The team felt that our schools should teach basic democratic civics to all students. Can we have students graduate from high school knowing little about the values that develop and sustain a community? To make progress on climate nationwide, through legislation and changes in our energy consumption, we need a democracy supporting these efforts. Climate action and a sound democracy go hand in hand.

Right now, our democracy is in peril. There is talk about a constitutional crisis. Many leaders no longer see a necessity to obey the law, tell the truth, or fight for the common good. Power and money are accumulating in the hands of the few. The topic of climate change is pushed aside while daily media reports about the latest political shenanigans capture our attention.

The good news is that climate awareness is growing and that people are increasingly making their voices heard. Ironically, social media have become an effective tool for getting the word out about both our imperiled democracy and our threatened planet.

These days, many people are discovering that working in groups is an effective way to make positive changes. We brainstorm together and learn from each other. As we overcome isolation, climate grief does not fade, but it does become manageable. We withstand failures better and have a chance to celebrate successes with our friends. When voices grow into a chorus, the sound travels further. Politicians listen, the country listens, the world listens.

This chapter presents voices of people who in various ways contribute to bringing people together.

Rachael Londer

When I was little, my mom spent time with me out in nature. She's a birder, so we would go on hikes and explore the Bay Area. She taught me about our Jewish principle, Tikkun Olam, of repairing the world to be a better place for all people and animals. My father was more interested in policy, so he liked going to city council meetings and making public comments. He inspired me to work on climate issues. He participated in organizations like the "Citizens Environmental Council" in Burlingame or "Sustainable San Mateo County." Because of him, I've always thought of having a career in sustainability and gravitated toward involvement-based local government.

He showed me how the local government has a lot of power. We can make a difference starting with our own block and our own city. That's where we can effect the most change. I believe my dad's mindset came from his parents, depression-era folks who thought about sustainability in terms of frugality—not using more than you need, thinking about your children, and wanting to provide for the future.

I went to meetings with my dad, but I didn't start making public comments until I was in college. I had taken an environmental science class the first year that San Mateo High School offered such a course. My teacher, Mr. Appleman, told us that you could make a career of seeking community feedback and making a community's vision come to life. He recommended that I attend Humboldt State University (now Cal Poly Humboldt).

At Cal Poly Humboldt, I did energy audits of our dorms, faculty offices, and classrooms. I volunteered for a nonprofit group, the Alliance to Save Energy. At the time, I thought I was having a huge impact on reducing energy consumption on campus. But really they were training me about small steps we could take, like how to install power strips and using different light bulbs. I learned how the university works as an institution and how we can do little things that may roll up into bigger policy and planning efforts.

While at Humboldt, I got involved with the county and made public comments related to the Community Choice Energy initiative that was starting from Redwood Coast Energy. It was right around the time they started to do outreach and to scope out what a Community Choice Aggregation (CCA) could look like in Humboldt County, purchasing electricity on behalf of residents and businesses in place of investor-owned utilities such as PG&E.

After receiving a Bachelor of Science in Environmental Policy from Cal Poly Humboldt, I moved back to San Mateo County and have been working in sustainability with local governments for the past 11 years. I am the sustainability manager for the city of

Menlo Park, which means I develop the city's climate action plan and see to its implementation. We have a bold goal of carbon neutrality by 2030. To reach that goal, we have many different strategies to reduce waste and to decarbonize by electrifying our buildings and transportation. We also work on positive carbon sequestration, looking at soil health and canopy coverage.

Electrification of buildings and transportation is more meaningful now than ever. In San Mateo County, we are lucky to have community-sourced clean electricity. Peninsula Clean Energy purchases electricity on behalf of every resident and business in San Mateo County. It's transmitted through PG&E's wires, but a local agency composed of a city council member from each jurisdiction in the county enables us to have local control of what type of electricity we purchase. Money gets back into the community through either lower rates or community programs like electric vehicle incentives or heat pump rebates.

What our county and city leadership did to take control of our electricity was transformational. It kept our dollars local and advanced electrification efforts. We accomplished the first step by getting access to clean electricity. Now we just have to power everything off of that clean electricity sourced from solar, wind, and some small hydro. So that's a huge development that really changed the game for sustainability. It enabled city staff to get funding for our electrification projects.

Peninsula Clean Energy is funding transition to solar across city facilities. They are helping us fund electric vehicle chargers and electrify low-income homes. They make it possible for us to offer rebates for low-income homeowners. We try to spread the word through the city's website, Menlo Park.gov. Not very many people access the information currently, so we are in the process of making it more user-friendly.

What resonates the most with folks is when we meet them on their block. We try to go to every community event in Menlo Park. I love the opportunity to speak with organizations or at the workshops they are hosting. I've been to many neighborhood associations and HOA meetings to talk about our work. I try to focus on positive things and a call to action. Instead of framing presentations around sustainability, I usually talk about ways we can save on our energy bills and ways we can have more comfortable homes by upgrading dilapidated electrical systems and appliances.

We need to bring the message to people where they're at, showing them options that are easy for them to implement. Tomorrow I'm going to the Belle Haven Community Climate Change team meeting. The Belle Haven neighborhood is east of 101 in Menlo Park. The team meets once a month. They've done really cool things, like distributing solar-powered battery generators to 16 homes. It was a pilot program. Those battery generators are available for checkout across the neighborhood, so residents can share these resources with each other. They can power their devices with solar energy and also get help in case of power emergencies or power outages.

Our Environmental Quality Commission meetings are another exciting part of climate work in Menlo Park. The city has appointed seven individuals to advise the city council on sustainability and climate-related matters. We meet once a month. Tonight we'll talk about improving our outreach around home electrification. We'll also look at potential building code changes we can make to encourage electrification.

Before launching a new program or starting a new initiative, I first think about how we can get everyone in the room together to talk about it. We build a coalition, seeking input that will shape which direction we'll go. Next, we may meet with technical experts to get all the right information, and then we present a proposal to the coalition.

At my workplace for the city of Menlo Park, our small sustainability team consists of three members concentrating on climate action, climate adaptation, and resiliency. Menlo Spark is a nonprofit working across San Mateo County, mostly on advocacy. They try to advance the amendments to building codes and look at how we can change our policies, ordinances, and regulations to require electrification.

My favorite part about local government action is that we have a lot of power. We have power over our streets, water, and utilities like energy and waste. I love bringing different folks together from community-based organizations to talk about solutions: climate advocates, business representatives, homeowners. Everyone brings a unique perspective and understanding of the issues and can offer valuable suggestions. My role is to bring all these different interests together so we can create a sound policy that works for everyone in the community and helps us along towards our climate goals.

We received a $4.5 million grant from the state of California. The Menlo Spark non-profit advocated for that money with Senator Josh Becker. To secure the funding, they shared a blueprint for how we could electrify homes in Belle Haven. Now the city can go full force, electrifying homes at no cost to income-eligible homeowners. With the first half of the funds, we're hoping to electrify 40 to 60 homes. That's a groundbreaking, tangible action, not only for reaching our climate goals but for improving the lives of some of our most vulnerable residents who live right off Highway 101 and are suffering from air pollution. They have also suffered from historic segregation that left them voiceless in our public process until recently.

The circumstances that led the community to be so disenfranchised and so disadvantaged has been well described. Now the city and community-based organizations are taking on this situation of environmental injustice. Efforts are underway to improve the air and water quality, access to healthy food, and access to clean and stable housing. A concrete plan was put in place last year. People are so excited. Upgrades are available for up to $50,000. Some people don't have electricity working in half of their homes. We are able to fix that while also giving them a new cooktop, a new dryer, a new water heater, and a new HVAC system, which also includes air conditioning, something most of our homes have around here.

So this progress is transformational for the neighborhood. We're hoping that with the 4.5 million dollars we can electrify 100 to 120 homes. Rebates are available through Peninsula Clean Energy for folks who aren't income-eligible and can't be served under this grant. We are hoping to mobilize contractors and just raise community awareness around electrification through this grant. We will see a big market transformation.

When I consider the future, I focus on actions locally because that's where I can be effective. I certainly am also concerned about the larger threats of climate change and national developments because the impacts on our community could be immense. Fortunately we haven't seen cuts to Peninsula Clean Energy's funding so far; nor has there been any rescission of grant funds. So for now, there has been no immediate threat to our work.

I spend a lot of time thinking about what sort of neighborhood I want to live in during the next 20, 30, 40 years. Then I work backward to what steps need to happen to make that a reality. I imagine living on a block with a community garden and a community pool and a library, a place where I know all my neighbors. We'd have a farmer's market, and I could bike to anything I could want with a five-minute hop on the train. And so I think about transit-centered communities, access to food, access to buildings that keep us comfortable on hotter and hotter days and that withstand or quickly recover from the more intense storms we'll have.

Right now I'm most focused on creating resilient homes, but that work will likely expand to all those other facets of a healthy community. I don't have much hope for improvements on the national or global scale. We're well underway to seeing the effects of climate change everywhere. People in the most vulnerable communities are already being displaced and suffering tremendously. But it helps to think about how we can become stronger together in the face of the changes to come. There are lots of opportunities to substantially improve the lives of our community members, like through affordable housing, lower utility bills, and all-electric appliances including amenities like air conditioning and electric-heated pools.

A lot can be done to make sure we don't go into a dystopian future, even though the climate is changing. In preparation for more extreme weather events, we can improve community infrastructures and form networks to support each other. Staying more local helps me remain hopeful. I find inspiration in our community volunteers. They create spaces where people can share ideas and resources. They support people as needed, whether it's checking in on someone on a stormy day, having coffee, laughing, or sharing a good book. All those little interactions and coalitions are much stronger than we give them credit for. They strengthen our social fabric, one initiative at a time.

For Further Reflection

- Have you taken an interest in local government? If so, how did you get involved?
- Are you aware of the sustainability efforts in your city?

- Which part of your community infrastructure needs reform? Are you willing to advocate for that?
- In one of the final paragraphs, Rachael presents her dream of a healthy local community. Which of her ideas resonates with you the most?

Folake Phillips

I moved here from Lagos, Nigeria, in July of 2017 to begin a family with my boyfriend. After we married, I got pregnant right away, so the first couple of years were spent preparing for the baby, looking for a bigger home, moving to Los Gatos, and taking care of my little one. I was just getting ready to meet more people when the COVID pandemic happened. To protect my son and the family, I didn't go out much or resume professional work until early January 2023.

The murder of George Floyd and the subsequent global Black Lives Matter protests motivated me to start a nonprofit focusing on education or social justice. The organization initially collected stories of individuals who were using the power of art—storytelling, music, dance—to create awareness. We curated community engagement spaces where people can gather to learn and share. Our emphasis was on topics that affect human rights, like racism, discrimination, xenophobia, domestic violence, and environmental degradation. The goal was to create awareness so that people could rise up and challenge abuses.

We still curate community engagement programs, bringing people together at the intersection of the three C's: color, class, and culture. We don't want there to be barriers around class. We want all beautiful shades of humanity to come together. That's what the Bay Area offers us because it's a microcosm of our world. And how do we begin to think locally and globally? It's making sure that more cultures are represented and more inclusion takes place in our local schools, government agencies, and community spaces.

For our events we use community spaces like the library, office space that our funders provide, a county property, or spaces from partner organizations like the Baha'i Center in Willow Glen, where on March 8th we'll celebrate the fifth anniversary of our Women in Action program. It's all about deepening community and finding a venue or a space that's affordable. Invitations to the event are put out on our social media page and our newsletter. We rely heavily on asking people to help us spread the word and to attend with a friend.

Word-of-mouth advertising is traditional and very effective. It's personal. When people hear about what has been put together and curated for them, it makes them excited to come and more likely to bring a family member or friend.

I read about Palo Alto's Fridays for Future program in a newsletter. When I connected with the group, I was impressed by how strongly the members felt about climate change and how ready they were to take action. I'm very attracted to groups that seek opportunities to connect as humans and who have a shared mission to protect people and the environment.

Santa Clara County of Public Health started what they call the "Climate Resilience Leadership Academy." We're part of a cohort of nine or ten organizations trying to engage all communities and make their outreach more equitable. I advocate for environmental stewardship and sustainability by way of personal lifestyle changes.

As I considered what members of my community might personally be affected by, I thought about the damage caused by PVC synthetic hair extensions. PVCs include very toxic chemicals that impact a person's health. Data suggests that they're used mostly by people of African ancestry. I won't go into the history of how societies impose a certain standard of beauty on people, although that's important. Racism is part of that also.

Anyway, at our event we learned about the Crown Act which is now the law in about 23 states but starting in California. We created awareness and asked people to download QR codes and sign a petition. The Crown Law supports and identifies with people of African ancestry and their natural hair, making sure that they're not penalized, lose their jobs, or face discrimination because of the way they wear their hair, whether it's natural, in dreadlocks, or braided.

But back to the toxic PVC synthetic hair extensions. They cause allergic reactions. Their carcinogenic chemicals are linked to respiratory issues for people who wear them. One of the professional braiders interviewed shared about a particular brand that she had used. While she was arranging a person's hair, her fingers started to swell. That had never happened to her before. Her talking about the harmful effects captured people's attention.

PVCs also harm the environment. They release pollution in the air because when heat is applied, the heat releases toxic fumes to the person wearing the extensions. The treated hairs clog up our waterways, and a documentary on YouTube revealed that over 14,000 tones of these synthetic hair extensions can be found in US landfills. So PVCs are bad for individuals and the environment as a whole.

As an organizer, I feel the best way to capture people's attention is to find things they can instantly relate to. Presentations really help. We partnered with stylists and salon owners. Our keynote speaker had a company on the East Coast that makes hypoallergenic non-PVC synthetic hair extensions and is now researching plant-based fiber options for hair. We had a panel to talk about use, recycle, reduce. We held a hands-on three-hour workshop. So people got engaged and we built a community. And all the braids used were from a hair company that provides healthier options.

We made a "throne," so to speak, using hair extensions all around it and putting Crown Act information next to it. We called it the "crown throne." People sat on the throne, felt good, and admired the art surrounding it. The chair was a found product from who knows where. It was amazing to bring people together for those conversations about the impact hair extensions can have on the person, the community, and the environment.

I am from Nigeria. Interestingly, the global south is disproportionately affected by climate change although it doesn't cause as much damage to the environment as the global north. The current US administration claims that climate change is a hoax and is pulling out aid, causing all sorts of harm and chaos globally. Nigeria is among the top 19 nations adversely affected by climate change.

We are a country of over 200 million. By 2050, we'll have the biggest youth population in the world. Creativity is bursting at the seams. But a lot more climate change awareness and education need to happen before we get to taking action. The average Nigerian is trying to get food on the table for the next day. So if we want to reach them by talking about the dangers of climate change, we must take a three-pronged approach: cultural, economic, and entrepreneurial.

Climate activists have to take the culture of the country into account when they educate about climate. They need to spread information that is understood. Lessons need to be applied to the situation on the ground. Take flooding, for example. In Nigeria, two years ago we experienced a huge flood. Businesses and organizations couldn't function. Flooding was so bad in certain regions in the southwest of Nigeria that people had huge property damage and even lost their lives. Communities came together, finding ways to support and provide for people, but the flood damaged much of the infrastructure.

When we talk about climate change in this context, we have to remember that poverty plays a big role. People are equipped to make repairs and grow their entrepreneurial spirit if they are sufficiently supported through local, national, or global funding. After a disaster, they are more likely to pay attention to climate-change information and awareness. And then funding empowers them to take action and advocate. But where there's poverty and people are thinking about where to find the next meal or clean water, it's really hard to share any sort of information and make it meaningful. So we have to look at the specific local situations.

We need to adapt what we teach to the Nigerian setting. Who are the leaders in the community? Who are the chiefs that we need to talk to so we can establish trust and build a connection? We're relational, communal people. It takes a whole village to make changes. So the approach has to show that we respect the culture.

My strength is the ability to connect people. I studied marketing management at Coventry University in England. I also worked with the corporate sector in Nigeria to implement corporate events and procure coverage. This background explains how I set up my nonprofit, using the power of community to bring people together for a purpose.

My value system was shaped by my family. I am the daughter of what in the US would be a congressman. My father served Nigerians at the national level. He was a pharmacist by profession. My Mom was a nurse, so both of my parents worked in the medical field. Even when he became an honorable member of the House of Representatives, he remained dedicated to his family. He always took care of his extended family, going back every so often to his hometown to work with people. He represented his constituents well and gave them a voice.

Both of my parents were compassionate. My Mom grew up Muslim and later became Christian. She welcomed people to her home. We always had people over, whether they were coming from different parts of Nigeria, Scotland, or the US. Our house was always open. I grew up with that hospitality, and it shaped me.

Personal values are a constant work in progress. I admit that I'm not perfect, but I focus on how I grew up, what is important, and what I believe in. My faith guides me to ask, "What can I do better, even if I'm not perfect?"

When I became a mother, I wanted to start nonprofit work, but as I looked at my little one, I wondered how to make the right choice. Then I told myself that eventually I would need to answer his question, "What did you do for society when all these things were happening?" I see everything now through the lens of a mother. What happens to children who look like him? What happens to children who do not have the opportunities that he has? These questions are guiding me in my work.

Today's youth are creating and changing the world. They are more energetic and master technology like no generation before them. I believe in the power of the youth for social change, transformation, and a cultural shift. They got my attention in 2020 during the Black Lives Matter movement. I saw youth who were angry and deeply disappointed. They felt let down. They felt the older generation had lied to them in many ways. I saw them standing by their friends, anticipating that they would get more harm. Many of their friends don't look like them but they stood together. And I thought, oh my goodness. All that passion, power, and emotion! Those are beautiful, important feelings that must be validated. Do the good work, be passionate, and keep it going. Find the right committed partners who can love your differences. We need to stay committed to humanity above self. I saw all of that in our youth.

Today's youth have so much faith in the future. How can we connect with them? Are we going to allow them to steward and guide us as we make changes? Will we surrender and relinquish power to them? They still respect their elders and they're still looking to us for wisdom based on experience, but they hold enormous power within themselves to help our world.

For Further Reflection

- How do you spread the word about climate change?
- Do you have places in your town where groups can gather to plan environmental action?
- Where does Folake see links between racism and environmental abuses?
- Have you used a product that later turned out to cause allergies or was declared carcinogenic? What do you do to protect yourself?
- What did you learn from Folake about dealing effectively with people of another culture?
- Talk about her invitation to relinquish power to youth as they lead us into the future.

Jordi Espin Marti

My parents own a small shop in Barcelona, Spain, called "Natural Remedies." They sell medicines derived from plants. When I was a kid, I learned about how you can avoid chemicals and be healed by plants instead. My family had this kind of alternative mindset, avoiding big consumerism or capitalism as a way of living. This impacted me.

Later I learned more about our threatened environment. Back in my last year of high school, we had to present a project. I proposed analyzing the quality of the water in a river close to Barcelona. My parents helped me gather samples of water from the spring feeding the river, from the middle part after some industrial regions, and from the mouth at the Mediterranean Sea. Then I analyzed with friends how the pH level and different ions were changing. So this helped me realize how water quality could have an impact on our society. It was a very basic study, but it motivated me to continue studying chemistry and read more scientific articles about environmental protection.

I enrolled as a student of chemistry at the University of Barcelona. For my PhD I studied a family of porous materials called metal-organic frameworks, the most porous materials discovered so far. When I finished my PhD, I wanted to use these materials for environmental applications. I found a research group in Switzerland focusing on both water and air purification for greenhouse gas removal. When I joined this group, I worked mainly on capturing carbon in the air. During the last five years, I have been leading a group of PhD and Master's degree students as a senior scientist to develop novel creative materials to improve CO2 capture.

I came to the Bay Area last September because my wife joined Stanford as a postdoctoral researcher in chemistry. I am currently exploring the job market, wanting to transition from purely academic laboratory research to employment in industry. I hope to have an impact on society by using my knowledge about carbon capture for potential new opportunities in corporations or startups in the field.

After I arrived, I joined Stanford's Bechtel Institute to socialize and meet new people. One of the employees there encouraged me to become a volunteer. So I began offering workshops for international students and their spouses on climate change. My first group met weekly last fall. I am now continuing for the second quarter. I lead the gathering by presenting slides and discussing climate concepts. I always encourage participants to ask questions and to talk about environmental practices in their own country. Yesterday a person from Korea described sustainability strategies that have been implemented. Other students from France, Japan, China, the Netherlands, and Argentina have shared in the past as well. It is important to learn from each other and hear about different experiences or points of view.

Yesterday I showed a slide related to a news item from last month. At Starbucks, some environmental activists tested how effective Starbucks was at recycling their plastic or paper cups. Were people throwing trash into the proper recycling bins inside the shop? They found out that of 40 cups only four were properly recycled. This behavior has quite an impact, because if people don't make an effort to separate their trash properly, all ends up in a landfill.

One person from South Korea shared that in Korean coffee shops, people are forbidden to use paper or plastic cups. You can either purchase reusable cups on-site or you can bring your own. If you bring your own, you may get a discount, so that's even better.

Recycling is an easy topic to talk about, but I also emphasize that sustainability encompasses the economy, the environment, and society as a whole. It's a very diverse topic. I tell my students how corporations can do better to report and curtail carbon emissions. We talk about the environmental impact and about what should be short-term goals for governments to reach net zero emissions in the future.

I tell them that many different greenhouse gases are accelerating the current global warming. CO2 or carbon emissions have the highest impact; but methane-chlorinated gases and nitrous oxide are also very damaging, with global warming potential that is higher than CO2. For instance, one molecule of methane has an impact 30 times bigger than just one molecule of CO2.

I remind them that in 2015, representatives from 196 countries signed the Paris Agreement, a document declaring that we will work together to reach net-zero greenhouse gas emissions by 2050. The idea was to reinforce or to pass legislation in each country to reduce and account for CO2 emissions, taking into account the different global economies and industrial levels. Second, each country would implement new regulations or offer taxes or carbon credits to lower carbon emissions in the near future. (Unfortunately, President Trump withdrew from the Paris Agreement during his first administration and again on the first day after his inauguration in 2025.)

For addressing climate change, every action counts. Many people tell me, "There's no need for me to do anything because if the big corporations or the governments don't take any action, I don't have any impact." I don't agree with that. Currently, around 8 billion people live on this planet. So I believe that if every one of us tries to contribute something very simple, we can achieve a lot.

As just one example, we can reduce the use of plastics, especially when shopping in supermarkets. Some markets have compostable bags for groceries. In others you can take fresh food without needing plastic bags or any packaging.

A lot of non-recycled waste ends up in the ocean. It's a big issue. We must ask ourselves, "What do I need? Can't I live without this item, this electrical device, this jacket? Do I really need to travel?" We should be conscious of every means of transportation we are taking. Also, every time we use our credit card, we are telling companies what we value. It's kind of voting on which product we want to have on the market. Once we figure out which companies are developing more sustainable approaches or practices, we will give more opportunities to these companies to thrive and then bring more of these products to our shops.

We also need to pay attention to how we get rid of things, especially electrical appliances and smartphones. Nothing like that should go in a landfill or just lie around home when it no longer works. Smartphones, televisions, or whatever electronic items we are throwing contain precious metals, sometimes even a high concentration of gold. So it's much easier to extract gold from all these appliances than mining it from virgin ore. Recycling electronic devices that we use every day is very important.

During the last five years, in addition to doing research, I have also enjoyed communicating, discussing with colleagues, providing new ideas, and brainstorming about new possibilities and their impact on society. I enjoy writing scientific papers, but I also love to listen to others and learn from them.

We have to stay optimistic, despite the many messages in the media that show how we have failed to deal with the climate crisis, solve the immigration problem, or change our lifestyle. We cannot stay depressed. Instead, we should trust that we can do better, learn from the mistakes in the past, and proceed differently in the future. Every small action counts.

I would love to see companies and governments aligning together to fight the big issue of climate change. In addition, each individual person should do their best, taking action as much as possible. We should capture carbon from the atmosphere that was emitted decades ago, scale up investments, and support technologies that can rapidly remove CO2 from the air. We need to make people conscious of the CO2 limits we can emit. We need to push for more legislation so that we can limit planetary warming to 1.5 or 2 degrees Celsius and avoid the melting of glaciers. Hopefully in 2050 we'll have more sustainable solutions for rebalancing equilibrium on our planet.

One interesting and affordable technology is in the process of being developed south of San Francisco. It's called Heirloom Carbon. They are using limestone as a starting material because it is rich in calcium carbonate. After a process of calcination, calcium carbonate is converted into calcium oxide, which then reverts to calcium carbonate when interacting with air. It absorbs the CO_2 instead of nitrogen and forms calcium carbonate again. And you can close the loop with such an inexpensive material as limestone. Renewable energies are used for the heating process.

So let's stay positive and make every action count. Be conscious every time you're going to buy something, because every time you ask for services, it will have an impact. It's up to us to choose which strategies or companies we support. And don't share the view that it is essentially already too late, because otherwise, people will just say, okay, I can sit back and not do anything.

Just look at what we have learned since the Industrial Revolution. There wasn't much environmental consciousness, only mainly economic growth. But consider the progress in our daily lives and in healthcare. Life expectancy has almost doubled in the last 200 years. Now it is time to think of what we can do better environmentally in a sustainable manner. That will be key in the next few years.

For Further Reflection

- Jordi describes how his parents influenced him to avoid consumerism. Which friend, teacher, or relative led you to be interested in climate action?
- What do you know about fruitful environmental activities in other countries?

- What was your reaction to the US withdrawal from the Paris Climate Agreement?
- Review the actions Jordi recommends for personal initiatives. Do you agree that every action counts, no matter how small?
- Do you think governments and companies will join to fight climate change? Explain why or why not.
- Which new technology have you heard about that will help the environment?

Elaine Salinger

I've always loved being outside in the natural world. I'm in awe of how amazing it is. When I am in nature, it grounds me; I lose myself in it. It's so much bigger and more beautiful than I could or want to be. When I was a child, I was fascinated with small animals, the way they are so delicate and so interwoven into the natural world. I have always remained in awe of them. All these animals and plants are being affected by climate change even though they did nothing to cause it. I understand that the world is not fair and that the Earth will survive no matter what happens, but I cringe at the thought that these precious beings are suffering. It's not fair. The same is true for people in the global south. It's not fair that they're suffering the consequences of our economy and the elitist way we live.

In my professional life, I used to be a veterinarian. When I retired, I had a bunch of things on my to-do list, but I kept honing in on climate change. The topic is huge, and I had no idea whether I was making a difference, but I wanted to do my part. So I began working two hours each day on habitat restoration. I first started in our own yard, then went further. We live on the edge of an open space. The hillside below our house blends into the open space. I remove the non-native invasive plants one by one. I discovered an amazing seed bank in the soil. After a while, all the native plants came back. It's incredible how our yard has been transformed! From spring to midsummer, three waves of colorful wildflowers fill the hillside.

I imagined that's how the hills along Interstate 280 used to be before the ranchers brought in bales of hay because they didn't think their cattle had enough food with the native perennial bunch grasses that stabilize soil and store carbon. The bales of hay brought in seeds and the invasive grasses and plants that took over. If that hadn't happened, we would have an abundance of wildflowers all along I-280.

I love sitting in the dirt, kneeling on it, and sometimes even lying on it sideways as I work on habitat restoration. I'm getting old, so it's more difficult not to change positions a lot, but the activity grounds and replenishes me enough to keep going. I continue to find peace in nature. I love sitting in the dirt and watching the spiders, ants, and millipedes. They are utterly fascinating to me.

About 30 years ago, we installed our first set of solar panels, and it cost us $25,000. It was important to me at that time to install them, and they are still working. Then, about 12 years ago, I learned about an offer to lease solar panels for eight years with no money up front. The solar panels would be installed on your roof, and you would pay for the installation over an eight-year lease. Because the energy from the sun is free, your bill would immediately go down. After eight years, you would have paid off the solar panels and then really start saving money. I thought, what a great deal! After having paid $25,000 for panels 30 years ago, this was a no-brainer, great for both the climate and your pocketbook. So I did a next-door post to about 500 homes. I figured everyone would want to do this.

Wrong. The only family that followed through were our immediate neighbors because they knew and trusted us, and they saw the solar panels going in. They talked to my husband, and you know, they went ahead with the installation. No one else did. That's when the light bulb went off in my head: People are not going to do the right thing, even when it's good for their pocketbook. You need to have legislation. The main reason I joined the Citizens Climate Lobby was my conviction that we need legislation to get people and businesses to do the right things. It's easy when you elect climate champions. But we need to work with whomever is in office to get legislation passed. And Citizens Climate Lobby is really good at teaching their members how to create political will for climate legislation, no matter who is elected.

Citizens Climate Lobby is a nonpartisan group that builds relationships with legislators, meeting them wherever they are on the spectrum of climate action capacity and then moving them along. Most people don't understand that we can turn the adverse effects of climate change around through legislative solutions. They also don't understand which solutions are most effective. They read about environmental crises, but it's hard for them to prioritize what laws are most needed.

And so I started doing presentations in schools because I knew students were upset about climate change. I talked with them in environmental science classes and environmental clubs about the solutions that really matter, and I emphasized that we can make positive changes. At the end of every presentation, I would ask students if they wanted to share their contact information with me so that I could reach out to them when a vote was coming up at a city council. If they wanted to write or had a project they were working on, I offered to mentor them and help to make it as effective as possible. Generally, after every presentation, four or five students would give me their names and contact information.

When COVID hit, all the environmental clubs closed down. The teachers were overwhelmed. They could barely manage their classes, let alone be the sponsor for an environmental club. So I reached out to all the students who had shared their contact information with me, and I said, "Do you want to start a weekly Zoom meeting on climate action? We could form our own environmental club for San Mateo County." About 36 of those students showed up at the first meeting. The club has continued. At this point, only six to ten students will attend every Saturday Zoom meeting, but they get a lot done. I still come to every one of those meetings, and I also encourage them to join Citizens Climate Lobby.

Recently, the group met with one of our supervisors for the county. They repeatedly go to city council members to discuss legislation. They organize an annual march and rally in the San Mateo Central Park. We invite both students and elected officials to speak, and we also have an open mic period for students who aren't part of the group. Then they march through downtown San Mateo with signs, and we usually have a marching band to accompany them.

Citizens Climate Lobby goes to Washington once a year for an annual two-day conference to talk about legislative priorities and opportunities. On the third day, we go to the Capitol and meet with our legislators behind closed doors. Those meetings have been amazingly informative and transforming because they have shown me how little most legislators understand about climate change. It's getting better now. But back in 2016, for example, I had a meeting with a legislator who said they were going to leave climate change up to the millennials. I started to cry, and that got their attention. It was hard to be an activist for climate back then because many legislators were just not on board. What's interesting, though, is that what they say behind closed doors now is completely different from what they say to the public. Many Republicans (but not all!) will publicly say that they don't believe climate change exists, that it's just a hoax; but behind closed doors, they admit that it's real and they know that we have to do something about it.

One thing I point out to youth is that it's important to have hope, that there is a path forward, and that we can fight climate change because we already know the solutions that will work. We must use all of them. We need legislation to make that happen, but we also have to understand that carbon pricing is by far the single most effective solution. The United States is the only developed country that doesn't have any price on carbon.

I do about 50 of these presentations every year, most of them in schools. I always start by apologizing to the students. I start by telling them that when I was their age, Congresspeople first started talking about climate change and how we needed to do something about it, but then they did nothing. Young people are going to inherit this massive problem that they would not have needed to address had the adults in the room done their job.

Then again, when you look at the general population, who has the time to do this work? The youth have a full-time job being students. Mid-lifers are out working, starting families, often even working two jobs. So who has the time? It's retired people. Unfortunately, most retired people only think about what they're going to have for dinner that night, watch TV, start hobbies, and travel around the world in airplanes.

When I talk to adults, I point out that one of my sources of joy and happiness is working on something bigger than myself, something that I can disappear into because it's a source of fulfillment and happiness. Then, once you have become a climate warrior, it's really important to understand how to maintain a balanced life. I've been doing climate action and advocacy for over five years, about 40 hours a week, and I am every bit as passionate now as I was when I started. But I have watched a lot of people who threw themselves into advocacy, and after two to three years, they just burned out and had to walk away. Activists need to figure out what brings them joy and how they can take care of their bodies so they can keep going. Because once you are a climate warrior, you are a rare and precious being. It's so important to maintain strength and balance. My own "secret sauce" is exercising for about two hours each day, being out in nature as much as possible, and working with fun and passionate people. My best friends are the other climate advocates I work with. And yes, I still do the laundry, the grocery shopping, and the cooking just like everyone else.

To think that we are removed from the effects of climate change and not connected to the natural world is a fallacy. It's a mirage that can evaporate in an instant, like it did for the people in the LA fires. One reason people don't take action is because no one has asked them or explained what needs to be done. They don't understand. They're confused. One thing I learned as a veterinarian is that when people are confused, they don't act. The steady barrage of negative news that the current administration keeps throwing at us has confused people and thrown us into inaction. Hopefully that will pass very soon, and there will be a big public outcry.

In terms of my own activism, in addition to working with students and Citizens Climate Lobby, I am a member of the San Mateo County Bike and Pedestrian Advisory Council. I was the chair for two years, but I stopped because it was so upsetting to receive emails and phone calls from cyclists and pedestrians who had been hurt by cars, and when we tried to get things done, absolutely nothing changed. That was discouraging and heavily weighed on me. But I remain a member of the advisory council. I also joined the Sierra Club executive committee last December because the Sierra Club is instrumental in endorsing state propositions and elected officials. Moreover, as the leader of the Citizens Climate

Lobby San Mateo County chapter, I send out all the calls to action. I meet with local elected officials, find out what bills they're working on at the state and local level, and then notify our chapter members to get them to call. This is how you build political will.

I'll share a little secret: Sometimes, we live in places where a candidate or elected official will come right out and deny climate science. But if you live in a place like I do, where that is taboo, you need to know what to look for on their website. If they're only talking about adaptation and not reducing our use of dirty energy, they are not a climate champion and do not deserve our vote. It's as plain as day when you know what to look for.

I sincerely hope that more people understand that our elected officials need to hear from us, that just voting is not sufficient. We need to understand which legislation is most effective, and then call and email to voice our opinions, over and over. Once is not enough. We need to join a group that has tools to make this fast, frequent, and easy. It is our collective action that works. And it's more fun to work with friends.

For Further Reflection

- ❖ How can you support the restoration of natural habitats in your area?
- ❖ What impressed you about Elaine's interaction with students?
- ❖ What is her "secret sauce" for staying resilient as a climate activist? What is yours?
- ❖ Elaine reveals her "little secrets" about effective legislative advocacy. What are they?

Reactions to Climate Change

Paintings by Elaine Salinger

Defense Against the Onslaught

Not Facing the Truth

Worries About the Future

Teri Whitehair

I've been working on environmental issues since the 1970s. The reason I get involved is that I care about the future. What motivates me is that I think about what we're going to leave for all future life. Back in the 70s we focused on conservation. Through the years, my focus has shifted from just conservation to a personal and collective commitment to make the kind of changes that will allow all forms of life to survive. I'm very much connected with nature, and I realize how important these changes are, not just for the people, but for all of life on the planet.

First of all, you have to recognize, especially now, the urgency of what it is that we're facing. Secondly, it's important to take it to the personal level as well. We have personal choices we can make. For example, my husband and I have been vegan for about ten years because we realized that what we eat and what we consume have an impact on the planet.

We transformed our older home into an all-electric house. We received an award for this from Peninsula Clean Energy. We have offered our home for tours so people can see what we did and be informed about steps toward going electric. And we have only one car. We walk whenever possible. Our son and daughter-in-law live in New York, but we've decided not to fly even though it's difficult to be so far apart. So it's personal choices like that—getting by with less and limiting the destructive impact on the planet.

The idea for the house tour started about two or three years ago. In the beginning, only a few people had started electrifying their homes, but then the movement grew and people began opening their homes, up and down the Peninsula, to encourage others to make changes. At one point, on a Saturday, people came to see what we had done. We went all out, preparing for this. We had posters about how much a particular installation costs, what rebates you got on it, and how much fossil fuel it saved on all the different appliances. So we walked the visitors through the various rooms and explained what we had installed.

Later on, real estate agents came to learn about the costs of making some of these changes. They had been told that it costs $250,000 to go all-electric, which is not true. For example, it's not true that you have to go higher up in amps in your house; instead, you can stay with 100 or 120. They had heard that installing a heat pump water heater was in the tens of thousands of dollars, where in fact it's around $7,000, and with rebates the cost is even lower. They thought doing the heating and cooling system was much too expensive because you had to refit your home. Well, our home was built in the 1940s, and we didn't have to refit it at all.

We also had the staff of the county's Office of Sustainability and their director come through. None of the staff had ever seen an all-electric home. We continued with the home tour for three years, bringing neighbors and people through. We invited students and high school students to see it, too. We've really made use of the home to educate members of the community.

In terms of becoming vegan, the original reason was my husband's heart attack about ten years ago. So at first, it was for health; but then the more we learned about the amount of water you save eating vegetables versus eating beef, the more we became convinced that it was the right decision. It takes about 2000 gallons of water per pound of beef versus maybe 400 gallons of water for a vegetable. You also have to figure in the cost of raising crops needed to feed the cows; all of that goes into the cost of producing one pound of meat. And water is a limited resource. So as we got further into it, we realized that living vegan was not only healthy but was saving the planet.

Elders Action Network had a campaign called "Take the Pledge." It asked people to commit to making changes that would help the climate, like buying habits, transportation, and so on. One of the options was becoming vegan. So I produced four videos showing how you could make an all-vegan breakfast, an entree, a side dish, and a dessert. During the videos, we also talked about all the fossil fuels this saves and the impact it has on climate. We posted the videos on YouTube, calling them "Cooking Vegan With Purpose." In each video I talk about not just how you do it, but why you should do it for the sake of the planet.

My husband and I are very active in our community because we realize that working with others has a much larger impact. We also are connected nationally with Elders Action Network, where I serve as vice president of the board. It is crucial to collaborate with members of the local community as well. So we started the San Mateo Climate Action Team. We also know the city council people, and we talk to them about reach codes. Reach codes spell out the requirements for office buildings and homes, and they are relevant for moving toward all-electric structures. The codes say that you have to reach a certain level by a specific point in time. We then advocate to the city council asking them to go beyond the reach codes as they seek to implement electrification. For instance, if they are working on a building code that prescribes switching to heat pump water heaters in ten years, we want them to go further than that. We urge them to reach that goal within five years instead.

I've given that example because often residents or contractors or developers will say, "Oh,we just can't afford that." The arguments usually come down to money. If there are countywide regulations that spell out when you have to change to a better system, then developers have to abide by them when they build. So we asked the council members to go further with the required building codes, and once they do that, everyone has to comply.

I have worked as a volunteer and community activist all my adult life. When our sons left for college, I decided to go back to professional work. I started out as an event planner for large events. I co-organized Beyond War, an award that was given out once a year in San Francisco. Then I engaged Eduard Shevernadzee to speak in San Francisco and arranged for Mikhail Gorbachev to speak the following year. After that, I was hired by the Gorbachev Foundation to organize the State of the World Forum. The five-day forum brought together leaders in political, social, nonprofit, religious, economic, and scientific fields along with other citizens. George W. Bush, Carl Sagan, and Ted Turner were some of the leaders who attended the forum. They would sit in a round table with ordinary citizens and talk about the issues that were important to them as they envisioned the future.

Next I went on to work with other organizations, such as the Institute for Research on Learning and IDG Executive Forums. Beyond War was an effort to change the way you think about resolving conflict. Even if you get rid of nuclear weapons, you still have to change people's mindset to get us beyond war. It involves not just the individual but all of us taking action.

Another thing that is very important to me is the understanding that we are interconnected with all life on this planet. We share the same DNA, even with the insects. Everything is linked.

My love of the outdoors and nature in general goes back to my childhood days with Uncle Pete and Aunt Mable. When Uncle Pete retired as a plumber, he bought a ranch in Northern California. He loved that ranch. My cousin Betsy and I spent a whole month in the summer there. We went out at 6 a.m. and milked the cows with Uncle Pete. I got to take the cream off the top of the milk from the cows. We collected the eggs. We checked on all the land and then we went back in and had another breakfast at eight o'clock. We'd ride our horses around. It was a 365-acre ranch, and it had such an impact on me! Uncle Pete gave me my own horse and I got to name it. I loved being outdoors, thriving in that environment. My Aunt Mabel grew all these fruits and vegetables and we would pick them and she would cook at night and make this big dinner for us.

These days, I maintain my connection to nature. Robert and I hike and walk and bike. We also camp at a lake in the summer. But my real strength is to reach out to people and to take initiatives. I'm a #7 on the Enneagram, so I'm definitely out there working with others. It's easy for me to talk to people. I'm also an organizer. I like to get a group together and build a team where everyone listens to each other, contributes ideas, and incorporates them. So that's where I get my excitement, when I see people working together, sharing what we know and what we do. I feel passionate about that.

Here's a story to illustrate it. When my oldest son was in middle school, every year the students put on a talent show that was not well attended. When I was president of the PTA, I initiated and organized a multi-cultural event with students and their families entitled "We Are One," mainly because my son had said to me that he had friends from Cambodia, Japan, Vietnam, and all over the world. We invited all the families to wear robes or other clothes from their country and to bring artifacts and a dish that is part of their heritage. To raise funds, we asked people to pay 25 cents to taste the dish. That event took place around 1989. Two months ago, I met with a city council member who said, "You know, every middle school in the city is putting on a cultural fair." I hadn't been aware of this cultural celebration continuing to this day in the city of San Mateo.

Here is a message I want to pass on: When you think of something important to do or are asked to do something, say yes first. A couple of times in my life, I said yes to an initiative or invitation although I had no idea how to accomplish the task or even knowing what I was really saying yes to, but it opened up so many other avenues. Saying yes to life, saying yes to the future, saying yes to all thriving—that's an important attitude to have. Think about it: When you first get married and you say, yes, you don't really know what you're getting into. But you say yes anyway.

Sometimes when we try things or we say yes, things don't work out. People get discouraged when they don't have the kind of success they expect. I run this social justice group in the Elders Action Network. We proposed forming a group to read and discuss banned books. We said, okay, we'll do it for three months and see how it works. Very few people showed up, and so we said, we're not going to continue this. I didn't feel bad. At times when something like that happens, I feel, "Oh, did something wrong." But when a proposal doesn't work out, it's not a failure. It's a learning experience.

For Further Reflection

- What projects is your city council working on currently? How can you move your council members toward climate-friendly solutions?
- What steps have you taken toward electrifying your home?
- Is becoming vegan an option you would consider? If not, why not?
- Which events from your childhood brought you closer to nature?
- Do you feel comfortable working with a team or would you rather tackle a project on your own?
- Can you think of a time when you said "Yes" to a project? How did it change you?

A Poem On Kinship With Life

Let us believe in the relatedness of all
that is: soil, willows, streams, and wind,
humans, whales, dandelions, dragonflies—
reality interconnected, all encompassing:
within, among, and far beyond,
yet intimately near in revelations
exceeding our widest hopes,
creating, sustaining, redeeming.

Let us believe that each embrace,
each helpful hand, will shape the rest
indelibly, although we may not realize
its impact at the time of our action;
for we are all knitted together
in a web without end; we all
stem from the same source,
share the same home.

Let us believe in possibilities of change
like tides that move the seas in rhythm
with broad and thinning faces of the moon.
Every seven years our body cells renew.
After each painful conflict we can seek
spontaneous smiles, new openings
beyond dead ends, a balm
that follows injuries forgiven.

In the name of solidarity and newness
and the creative power
in all of us.

Part 6

Guardians of the Earth

Voices:

Robert Whitehair
Ellie Cohen
Gerald Bernstein
Bill Myers
Cliff Kindy

Ellie Cohen is the CEO of The Climate Center, an environment and clean energy policy nonprofit working for accelerated, equitable climate action in California. On February 13, 2025, she gave a talk for Elders Climate Action NorCal, titled "As Goes California, so Goes the World." She described our climate crisis this way:

The current climate goals that most countries are working towards are bold, but they are inadequate for where we are today. They were established by the United Nations Intergovernmental Panel on Climate Change in 2018. The goal was to essentially cut emissions in half by 2030 and draw down as much as one trillion tons of past climate pollution from the atmosphere to get us to carbon neutrality. But even that's not enough. We need to get to net negative, removing more past climate pollution than we're currently emitting. Worse yet, emissions will likely even increase between now and 2030, not decrease. So, as the UN General Secretary has said, it's "code red for humanity."

In 2024, we experienced record carbon dioxide emissions globally. It was our first calendar year at 1.5C global heating. We are now in a La Nina phase, which usually means cooler ocean conditions; but warming didn't slow down. In fact, it climbed to 1.75 C degrees over the industrial era average from 1850 to 1900. This indicates that perhaps we're already passing some global tipping points that lead to potentially irreversible changes in climate.

Part of the problem is a surge in methane emissions. Cutting methane is one of the fastest ways to slow down this onslaught of climate change. As the Greenland ice sheet melts faster, the Atlantic current that transfers heat from the southern hemisphere slows down. It is now expected to stop completely in roughly 20 years if we continue climate pollution as usual. Once that current stops, dramatic changes will happen. All that heat will stay in the southern hemisphere and accelerate the breakdown of the Amazon rainforest, the "lungs" of our planet. It will also speed up the melting of the West Antarctic ice sheet and the permafrost regions, creating a devastating feedback loop.

Think of global warming as a blanket around the Earth, much thicker than the thin blanket we need to sustain life as we know it. Even a little increase in thickness makes it so much warmer and throws all our Earth systems out of balance. The oceans are holding a huge amount of this heat and drive extreme weather events. Last year the ocean temperatures were the warmest on record.

In the US, the trend of producing more and more oil has been going on for some time. We are number one in the world for plans for new oil and gas extraction through 2050. We are the greatest climate polluter over time in history. As Californians, we need to take responsibility to turn this around. We have the world's fifth-largest economy.

Globally, $7 trillion in subsidies were given to oil and gas industries just in 2023. Lobbyists from Big Oil and Big Utility are talking with legislators every single day. And of course, the new president has withdrawn the United States from the Paris Agreement, joining Iran, Libya, and Yemen as our bedfellows. It's a very sad time.

Fossil fuel air pollution kills millions every year. Over 7 million Californians live within a mile of active oil and gas wells, exposed to pollution, toxins, and carcinogens. And extreme heat is the biggest single killer from climate impacts, especially for farm workers and other outdoor workers.

The bottom line is that every tenth of a degree of warming that we prevent makes a huge difference for people, for wildlife, and for our future.

Robert Whitehair

I have done much work with electrification, but that effort is not broad enough for me. My wife Teri and I are both members of the Elders Action Network, and we both participate in Regenerative Elders workshops. Recently I got involved in a nine-week course called "Resilience and Acceptance in the Face of Collapse." We talk about the situation in which we find ourselves and about what it would take to get back to acting with a kinship view of the planet, realizing that all created life is interconnected, interdependent, and interrelated. Following the writings of Norma Wong, Regenerative Elders look over the horizon to what's coming. Most of us are likely not going to experience most of the global climate disasters because that horizon is probably, you know, 20 or 30 years away, but the collapse has already started and will be total.

The reason I don't like to focus solely on electrification as the center of my activities is that by itself it will not save the planet. I approach electrification as an appropriate way to have a kinship view of the planet.

I have become involved in the process of acceptance of resilience in the face of collapse. Because it's real. It's true. It's not doom and gloom; rather, it's expressing acceptance of the path we're on, which is freeing. Electrification won't get us out of the muddle. It will reduce the impact on the planet, and it's the right thing to do moving forward—but it's still part of the exploitative, extractive state of affairs that we're in. My journey over the last two or three years has led me to accept the kinship view, the total acceptance of where we're going and what we have to do.

I believe that elders have a tremendous role to play, given that we don't need to go out every day to work at a paid job. We don't have to worry about finances. Teri and I are not rich, but we don't have to worry ever again as long as we have food and shelter. So first we acknowledge that we have a certain amount of freedom as elders. The next step is to personally transform ourselves to be in touch with what is happening in the world. And the third step is to take action based on that transformation. I accept my relationship to the planet. I accept that I am not in control but that I have an obligation to take action to improve the lives of every single living creature on the planet.

I'm channeling a little bit of Jimmy Carter. That was Jimmy Carter's mantra. Every day, do everything I can to make my life and the lives of those around me as good as possible. Jimmy Carter was the one who got me started being an activist. When he talked about limits to growth, he created the conditions and the atmosphere for people to try to live with less. I took a course back in my college days where toward the very end we learned about the total production of "stuff" in the world. I forgot what it's called. Anyway, I learned that there are limits to growth, that resources are not infinite. They are finite. And there is a resulting cause and effect.

Accepting the climate disaster does not mean I accept it in the sense that there's nothing we can do about it. It's not doom and gloom. It's accepting reality, and helping others to accept that reality. A community means everybody has their own niche. Everybody has their own value. And we all contribute. Mine happens to be electrification. That's my niche. But it operates within the bigger picture.

Norma Wong talks about "over the horizon." She's indigenous Hawaiian and Chinese, and what she's saying is that elders are not going to be here forever. I probably have six or seven years left. I have a lot of maladies and problems. It's the people "over the horizon," my grown sons and my grandkids, who are going to be facing the climate future. I'm probably not going to be here, but I can warn them, help them, show them the value of people working together in community.

During college, I demonstrated against the war, even though I was in the Marine Corps Reserves. That got tricky. I went into the Marine Corps Reserves to stay out of Vietnam because I was sure I would get killed. I didn't have the power of the convictions to be a conscientious objector at the time. Maybe 20 years later I would have become a conscientious objector.

In the late 1970s I gave a talk on limits to growth. I'll never forget that talk at the Palm Springs, California, Board of Realtors meeting. How confrontive, to talk to realtors about limits, people who sell and want to create more development! Anyway, I said that there are limits to growth and that we cannot continue to grow endlessly on this planet.

Carter already pointed that out in the 70s. He put solar panels on the White House roof. Reagan, by the way, took them down immediately. Just like Trump is going to take down initiatives that Biden started. He's going to reverse everything. He said "drill, baby, drill," pushing for oil consumption before and after his inauguration. Remember the movie *Clockwork Orange*? At the very end, the main character is forced to hold his eyelids open with straws and watch all the terrible things that happened during his lifetime. I feel that way when I listen to Trump. It's very, very, very, sobering. You know, it's serious. The easiest way to work with him is just to think of the opposite of what he says. If he says he's going to clean up this or that, he's going to screw it up. If he promises justice, there's going to be injustice. It's going to be exactly the opposite. He has a very dark side.

Enough said. Back to the limits on growth. Teri and I got involved in the Creative Initiative Foundation. Harry Rathbun, a law professor at Stanford University, founded the organization along with his wife Emelia.

Harry Rathbun said something that stuck with me, and is formative of my beliefs as an elder: "The best test for truth is experience."

At this point in my life, as an elder, I've had lots of experiences. I was born poor but at some point I told myself that I wasn't going to live a poor life. So I went to work trying to make money, surrounded by people who are very rich. But when I retired, I got involved in limits to growth, accepting reality, and making personal changes.

We use no so-called "natural" gas in the house. We removed the gas meter, and I personally pulled all the gas pipes out. We've been 100% electrifying for three years now. We use 24 solar panels. We have no electricity bill, and our utility bill comes to only $10 a month.

Diane Bailey from Menlo Park started an initiative called "Menlo Spark," a campaign for fossil-free buildings in Silicon Valley. Her goal was to get rid of fossil fuels, period. Because of that goal and because of her actions, San Mateo got a $4 million grant from the state of California to build electrified low-income housing. They are in the middle of that effort now in Menlo Park. Peninsula Clean Energy is administering the funds. Most of the money will go to low-income houses in Belle Haven and other parts of San Mateo County.

City councils unfortunately tend not to look at the big picture about climate, but we shouldn't expect any government to solve our problem. We're going to have to solve it through our own personal transformation and our own lifestyle, through the way we live on the planet. And that includes the way we drive, the way we eat, the way we modify our homes, the way we do everything.

To make change is to take opportunities as they present themselves. For example, tomorrow we'll meet with advocates in the city of San Bruno who are going to speak next Friday at their city council's goal-setting session. The way to get people motivated is to pay attention and look for opportunities to give to others. I was saying earlier that we're all a community. We all have a niche in it. To make things happen, we need to help others build their own communities. I can't build a community for other people, but I can show them what I think works in some places. Then they can adapt that for themselves in other locations.

Let me tell you the story of a local climate action team here in San Mateo that illustrates how important it is to work with others. This wonderful woman, Michelle Hudson, came to me one day and said, "I understand you know about electrification. I'd like to work on that as my response to the climate crisis." So she and I got together. She has some really good strengths in some areas, I have some in others, and we run ideas by each other. Teamwork and community are important. We should demonstrate that in everything we do.

The next four years under the current administration are going to be tough. We need emotionally to accept what's coming and not try to resist as the only method of response. Work with others. Look over that horizon that Norman Wong talks about. Perhaps we have about a year where things will be glorious for some in terms of wealth accumulation. But the administration cannot handle how to govern. They'll do a terrible job at it. Sooner than later, there will be a crisis. So be prepared. And love your neighbors. Help them as much as you can, every single day.

What's causing our climate predicament is our lack of a relationship to the planet. What I'm thinking about is the cannibalism of capitalism. It's the greed of cannibalistic capitalism and the desire for money. As a people, we're always pursuing more and more riches, more and more things. That's the real problem. It's not Donald Trump or the Republican Party. It's each one of us individually pursuing capitalistic greedy cannibalism. Capitalism has to grow about 2.5% to 3% a year to function. In order to make that happen, we're eating all our resources as fast as we can, and we're fouling the air.

That's what we need to be aware of. Learn about the issues and take to heart how we can address them, without looking to the government for solutions.

I feel hopeful when I get into a group of people who talk about that. It's liberating. On the other hand, in a group that spends time only griping about Trump, the sense of freedom falls apart pretty quickly. It morphs into anger and resentment and hatred, just like the anger and resentment and hatred that he himself feels. Donald Trump never came to grips with this dark side. His father taught him how powerful anger is. If you don't connect to a community that works for the common good, you'll fall into anger. I'm in two or three groups now who meet weekly to talk about the predicament we're in, but we base it on reality and encourage each other. That gives me hope.

For Further Reflection

- ❖ Do you share Robert's kinship view of life: being interconnected, interdependent, and interrelated? Can you think of examples for each one?
- ❖ What does "acceptance" mean for you as you face the climate future?
- ❖ Robert's "niche" in addressing climate change is electrification. What is yours?
- ❖ How do you deal with the fact that our planet's resources are limited? What are you willing to give up?

Ellie Cohen

Ellie is a leader in catalyzing cross-boundary, collaborative, and just solutions to climate change and environmental degradation. The National Women's History Project named her one of 100 women taking the lead to save our planet. She challenges all of us to take bold action for a climate-safe future. Below is her eco-biography, followed by some more comments from her ECA NorCal presentation, specifically referring to California and her work at the Climate Center.

In 7th grade, my goal in life was to bring Russians and Americans together through our common love of nature. For college, I applied only to schools that had Russian and forestry programs. It was in college that I learned forestry wasn't for me, since in those days it was about cutting down as many trees as possible. I majored in Botany and took every class I could that would get me outdoors. I had a great time! I got a job doing butterfly ecology research that brought me out to California, where I had been accepted to get a PhD at UC Davis, but then a trip to Brazil changed my life's trajectory.

While in high school, I went with AFS (American Field Service) as an exchange student to Denmark for a summer. My family became involved in the program regionally. It was transformative, opening our eyes and hearts to the world. After college graduation, I visited my high school AFS sister who had lived with my family when I was a senior. We traveled around Brazil, where we saw poverty on a scale I had never experienced before. Upon my return, I decided not to get my PhD and to do community organizing instead.

In my 20s I had the honor of befriending Fred Ross, then 75 years old, the man who had inspired and trained

Cesar Chavez in community organizing. I convinced him to teach me the house meeting method that the UFW had used so successfully to get over 20 million Americans to boycott table grapes and iceberg lettuce, resulting in major improvements for farmworkers' work conditions here in California. We successfully used house meetings to target one member of Congress, mobilizing hundreds of people in his district to change his vote and oppose the deployment of nuclear missiles in what was then West Germany.

A few years later, we used this same house-meeting strategy to change enough votes in Congress to stop US military support of the Contras in Nicaragua, effectively bringing that war to an end. A major donor paid for about 12 of us to visit the country soon after. We were able to meet with people in small villages that had been under attack just two weeks earlier. It was quite moving. The success of the entire effort made me believe in the power of organizing and the promise of democracy!

In 1999, I became CEO at the Point Reyes Bird Observatory (now Point Blue). I worked there for 20 years, fulfilling much of what I had dreamed of doing in my life. I am proud of the collaborative accomplishments during my tenure. The organization grew fivefold to a workforce of 200 and a $14m budget. We secured $100 million of conservation investments in 2+ million acres of ag land (rangelands, croplands, creeks, forests) for water, wildlife, carbon, and people; and we guided 95% of urbanized CA coast (70+ agencies) in preparing for sea level rise and extreme storms. But having seen the impacts of climate change on birds and other wildlife first hand, I decided that I needed to work full time addressing the climate crisis. In 2019, I became CEO of The Climate Center.

Collaborative accomplishments under my leadership in California include passing a first-of-its-kind natural climate solutions law and securing 81 state targets for 2030, 2038 and 2045 covering more than 50 million acres in California; securing $3 billion for decentralized clean energy such as rooftop solar, microgrids and batteries; and, enacting a nation-leading bidirectional EV charging law to allow the enormous collective energy storage of EV batteries to make electricity in the state more affordable, reliable and resilient.

In California, we have what one scientist called weather whiplash. We have these super wet periods in winter, followed by super dry periods. Population growth and land management changes have led to about a third of the people in California living in areas subject to wildfires. Put all that together and it's a recipe for disaster. Just since January 1st of this year, we've had 359 wildfires in California. The LA fires were horrific and shocking. It's not just the areas where the houses burned down. It's also the exposure to toxic smoke and ash.

All of us have been exposed to wildfire smoke across California. Between 2008 and 2018, wildfire smoke killed over 50,000 Californians. The effect of wildfires during that time led to $432 billion in health expenses. We know that the more we invest now in reducing greenhouse gas emissions, the more lives and dollars we will save down the road.

The nation and the world look to California to lead the way to a climate-safe future. However, California is also the third-largest refinery state in the country and the sixth-largest producer of crude oil. Let that sink in for a minute. We are an environmental state, but we have this contradiction. We are the second largest consumer of petroleum products and the biggest car market in the country. We have a big challenge here. When we want to pass aggressive climate laws commensurate with the rapidly worsening climate reality, it's extremely hard. We have some powerful moneyed interest groups lobbying for their short-term profit even though they know and have known for decades the deadly nature of ongoing fossil fuel pollution.

California's current legal goal is to cut emissions 40% by 2030, which is way too low. We know we can and must do much more, but it requires political will. Big Oil is spending enormous amounts of money to stop us. In a way, that's a good sign. It means we're making progress.

At the Climate Center we developed a campaign called "Climate Safe California." It's about decarbonizing our economy as quickly as possible and moving to clean, affordable, and decentralized energy to ensure a resilient, vibrant future for everyone. The prescription for getting to a stable climate is not only to cut emissions dramatically but also to draw down past emissions upwards of a trillion tons.

We have several ways to get to that. At the Climate Center, everything we do is guided by three principles:

1. Stay based on the latest science.
2. Prioritize climate justice and equal access to climate solutions.
3. Ensure a just transition for the workers, their families, and communities that are currently dependent on the fossil fuel industry.

We have to triple the amount of clean energy deployed, and we need about eight times more battery storage to reach our goal. That means investing big in local clean energy resources. We want to draw down past climate pollution through nature-based strategies as well as technological carbon dioxide removal. And we need to fund these efforts at the scale needed.

No single organization can do this alone. We have to work together. We'll never have the money that Big Oil or Big Utility have to lobby. But what we do have is people power. We are making a difference, and we just have to keep working together to move the needle more and more.

So I want to share with you some of our progress. The Climate Center played a lead role in establishing community choice energy (community choice aggregation or CCAs). Marin Clean Energy, the state's first CCA, was established in 2010. CCAs are electricity-providing entities overseen by local elected officials, as opposed to PG&E and other for-profit, investor-owned utilities, or IOUs. Sonoma Clean Power, the second CCA in California, was established in 2014 thanks in part to the Climate Center's collaborative leadership. The Climate Center then helped more communities around the state to establish CCAs. Today, 23 CCAs serve about 80 to 90% greenhouse-gas-free electricity to over 13 million Californians—a major success story about speed and scale decarbonization.

At The Climate Center, we work to electrify everything. That requires creating a decentralized integrated power grid. We have a vision of "The Grid For the Future" that is safe, affordable, resilient, and reliable, prioritizing frontline communities that suffer most from air and water pollution, power outages, and lack of access to climate solutions. We've had some successes over the past several years. We catalyzed a new state community energy resilience initiative that started with $170 million in federal funds. We secured $3 billion in the state budget over the past few years for residential solar and storage, with a significant percentage of these funds dedicated to working-class households.

We have also been leaders in "batteries on wheels"—EVs with bidirectional charging capabilities that can help us store intermittent energy like solar and wind and use it when needed for outages or when there is a lot of demand, like during a heatwave. Then we don't have to use polluting natural gas or methane gas fossil-fuel plants that are often located in low-income communities. We passed Senate Bill 59 (Skinner) last year, authorizing the state to require classes of electric vehicles to be bidirectional.

In case of a power failure, I can now plug my house into my car. The batteries in my car are equal to more than five Tesla Powerwalls. We have an enormous amount of EV battery storage potential with two million electric passenger vehicles on the road in California today. Roughly, that means 20 gigawatts of storage in those batteries. Compare that to the highest ever energy demand in California, which was 52 gigawatts. Many studies show that utilizing even a small portion of EV batteries saves billions of dollars, makes electricity more affordable, ensures cleaner air as we reduce reliance on polluting methane gas plants, and helps speed up our urgently needed transition to a clean energy future.

In 2022 we also passed a nation-leading natural climate solutions law. It required that California establish legal goals for healthy soils and protected habitats including agricultural lands and urban greening. Healthy soils can sequester carbon back from the atmosphere into the ground. When it rains, instead of all the water running off, healthy soils and habitats can absorb it, slow down flood waters, and replenish groundwater. Healthy soils and habitats support food security and wildlife. Moreover, healthy ecosystems reduce pollution.

We worked another year and a half on that with many partners and state agencies. The first-in-the-nation targets came out last April, covering more than half of California, with goals for 2030, 2038, and 2045.

We're also working with partners to rapidly scale up climate-friendly agriculture. We restored some of the state's historic five-year climate budget as California faced a budget deficit, and we saved hundreds of millions of dollars for clean transportation, clean energy, and natural climate solutions. Working with many partners, we helped to pass the $10 billion climate bond out of the legislature and get it placed on the ballot for November 2024 as Proposition 4. It passed overwhelmingly with 60% of the vote. California's voters care about the climate crisis and are willing to pay for cutting emissions and reducing impacts on their communities.

We need to do more, of course. We're working on cutting emissions and phasing out polluting fuels. The

LA fires are expected to cost at least $250 billion, yet the total yearly budget for California is around $330 billion. So we know we have to find other sources of money. Big Oil keeps making billions in profits as they spend record amounts on lobbying against clean air, clean water, and a healthy future for us all. We are holding many organizations accountable to make sure that they pay for the damages their products have caused—what we call "make polluters pay." And we are working to ensure that the state is not giving away tax breaks or other financial benefits to Big Oil.

We are working with a coalition to address one of the biggest gaps in state energy policy: the lack of a plan for phasing out oil and gas production, extraction, and refinement in the state.

The single biggest source of emissions reductions in California is our cap-and-trade program. Industries buy allowances within the cap on total emissions and can sell them to each other. Right now, the cap is not tight enough, and the cost of carbon is too low, $40 per ton when it should be $200 or more. One study has shown that eliminating fossil fuel pollution in buildings and transportation in California would save $44 billion a year. It would also improve health and avoid 5,000 premature deaths per year. We must invest big now for a healthy, vibrant future for all.

Our goal at the Climate Center is to advance accelerated equitable policy and build partnerships to move the needle in Sacramento, inspiring other states and countries to do the same.

Millions and millions of people now have experienced some form of extreme weather because of climate change. We must now turn this into an opportunity to enact the bold climate policy needed. Everyone can make a difference—at home, in our neighborhoods, in our cities and counties, and beyond—to ensure positive tipping points in society. Working together we can and will ensure a climate-safe future for everyone!

For Further Reflection

Imagine yourself working with climate activists. What image resonates most with you?

- ❖ Stones thrown into a lake, forming ripples that spread.
- ❖ Instruments of an orchestra, each producing its own melody but blending with the ensemble for a common goal.
- ❖ Mosaic pieces that are broken by themselves but work together to form a beautiful and cohesive picture.
- ❖ Kaleidoscopes turning, changing, creating ever new patterns—but continuing to point to a source of light.
- ❖ A non-violent army moving forward and building powerful alliances: followers and leaders, the loud and the quiet, the young and the old, those who fight at the front lines and those who provide support.

Gerald Bernstein

Two strands, my professional life and my faith, initially evolved separately but then came together and started reinforcing each other. Let me begin with the career part. I have a BS degree in Aeronautics and am a private pilot, though most of my early professional work was in transportation and city planning. I moved to California to attend Stanford University's Civil Engineering Master's program in Planning and Environment. While at Stanford, I began working with a faculty member doing consulting for different airlines. This led to a subsequent position in the transportation program at SRI International in Menlo Park, a nonprofit research institute that collaborates across disciplines to bring science innovations to the marketplace. I started at a junior professional level, then rose to senior professional, then became a manager and later a director of the transportation program, which for a time combined with the environmental program. It was a great place to work and a great career; I earned numerous achievement awards during my sixteen years there. In 1995 I spun off a consulting partnership with former colleagues, maintaining many of our same clients.

But by the late 90s, I began to ask myself, "What are we doing with this business outlook?" We were looking ahead 10, sometimes 20 years, developing forecasts and seeking to expand the aircraft industry indefinitely. I started wondering how we could grow the economy endlessly on a finite planet. Doubts started creeping in about the unlimited economic growth model.

By 2000, the California energy crisis helped me recognize that energy sourcing and consumption were an environmental challenge. I started paying attention to the UN panels' reporting about CO2 emissions and climate change. I knew the aviation industry would continue to grow at a rapid rate, faster than GDP; it might need to take land from food production or cut forests to make sustainable aviation fuels. But I wanted to move from carbon production to carbon reduction.

I had children by then, which may have influenced my decision to look for another, environmentally-friendlier career. An acquaintance at City College of San Francisco (CCSF) managed a program called "Advanced Transportation Technology." She was going on sabbatical leave and asked me to watch over the program for a year. (CCSF is part of the California Community College system.) The program's goals were unclear to me as similar programs in other parts of the state had different focus areas (e.g., aviation, auto training, smog reduction). But they did appear to share environmentally favorable aspirations.

This career move involved a pay cut, but it was a step in the right direction because I started working with people concerned about carbon reduction. Earlier, in the 1990s, California had enacted an initiative for auto manufacturers to sell zero-emission (electric) vehicles. The problem was that at that time, people didn't buy the vehicles because they were so inadequate. But at CCSF in the mid-2000s, several people remained interested in electric cars. One of them had converted a Volkswagen into an all-electric Rabbit. In those days, you could buy a kit, put in a lot of batteries, and make this conversion. More practically, they were trying to introduce hybrid vehicle training into the traditional automotive technician training program.

When my friend returned at the end of her sabbatical year, she decided to move to another program and said, "Keep it." I ended up spending the next ten years at CCSF.

The automotive faculty wanted to move into hybrid technician training. But to teach how to repair a hybrid, you need training courses for faculty, tools, and properly equipped facilities. And you need a car to work on. But the auto program didn't have funds to buy a car because technical education programs are always short on funds.

I said, "Let me try something." I called my auto insurance agency and asked, "If you guys ever get a wrecked, rear-ended Prius, let me know." We needed a car that had been rear-ended because the front is where the engine, electric motor, and transmission are located. Four months later, I got a phone call, "Your car's in Fremont." So now they had their car. You couldn't drive it because it had been written off as a wreck, but the front of it was still whole and usable for training.

With this experience and faculty support, I applied for and won a three-year US Department of Energy grant to expand hybrid vehicle technician training to automotive programs in other community colleges throughout the Bay Area and, later, statewide. This funding enabled the colleges to expand their automotive technician programs from the gas engine cars of the 70s to the hybrids of the 2000s.

That was my first big step. Next I focused on alternative energy education. I partnered with a local solar photovoltaic (PV) installer to build a mock roof and offer workshops to the public explaining the hows and benefits of residential solar electric systems. (These are the photovoltaic panels you now see on rooftops everywhere.) Solar thermal is what people might have for swimming pools or hot water; PV is the electric version.

By way of background, the first push for solar came in the 1970s when President Carter put solar thermal panels on the White House roof. Those were hot water panels that Ronald Reagan then promptly removed. The industry had problems because every plumber claimed to know how to install the panels when in fact they didn't. They weren't used to dealing with the high temperatures, and a lot of these systems leaked. (I swam at an outdoor pool in the 1980s that had hot water panels that were never used because they leaked.) The panels were a great idea, but people didn't install them correctly.

Decades later as the solar PV industry began to form, participants said, "We're not going to get the black eye that the solar thermal people got. We're going to develop a training program so that our professionals can say, 'I have the training and a certificate to document that I'm knowledgeable about the proper design, installation, and maintenance of a solar electrical system.'"

A community college faculty member in Concord had taken the national exam to get certified and had developed a course about PV installation. Soon other faculty said they were interested in teaching this too. He prepared a training course for a group of us, including faculty from the electrical programs at several Bay Area colleges. Many of us took the exam and passed.

Soon thereafter, the Department of Energy created a five-year, nationwide faculty PV training program, with the California Community College Chancellor's Office selected as one of the nine participants. When the state got the five-year grant, the college administration started searching for a person to run it. They looked at me and said, "You're in the transportation and energy program; you've got the credentials for proper and safe solar installations; you know the standards; you run this." This is how I ended up setting up PV installer programs at about 50 community colleges statewide.

After ten years at CCSF, I was getting close to retirement. I wanted to continue working on clean energy, but less than full-time. When I looked for possibilities, I found the Interfaith Power and Light (IPL) organization; I thought maybe I could help with fundraising since I'm used to writing proposals to corporate and government entities. A board member I knew from the San Francisco Zen Center suggested that I talk to their director, who said, "Sure, we could use a volunteer." I didn't have any idea what I would do, nor did she.

However, shortly after I joined, IPL received a grant from the San Francisco Department of Environment to assist churches in reducing emissions through energy efficiency measures or solar PV installations. We had an intern who did a great job working on this. I worked with him, and when he left, I said, "Let's do this statewide, even nationwide." I developed a program to help congregations evaluate whether a PV investment would be financially and environmentally beneficial before they involved an installer and developed a formal proposal.

This is an interfaith program. I've worked with churches, synagogues, and mosques. The Inflation Reduction Act of 2022 (IRA) spurred a lot of interest in going electric. It includes a tax credit that lets someone installing a solar PV system deduct 30% of the cost. But since non-profits, including houses of worship, wouldn't benefit from such a credit because they don't pay taxes, IRA offers a 30% rebate to these organizations. In other words, instead of a tax credit, you get a check. Interest was high. Unfortunately now, with the new administration, we don't know what will happen.

I am pleased to have spent these later years of my professional life in work that protects the environment by reducing carbon emissions. The outdoors has always been a comfortable and friendly place for me. I grew up in a small Maine city. My family was Jewish, but I was introduced to other faiths early on. My dad described himself as a "church tramp." He was brought up Orthodox Jewish but completely reacted against this once he left home. When I was about ten years old, he started taking me with him to a different church each Sunday to show me the different approaches to faith that people have. I continued studying in the Jewish tradition and had my bar mitzvah at 13, but after that I enrolled in the Unitarian church for two years as there was no follow-on activity for Jewish youth. In the worship services I attended, it struck me there was always a lot of talking—sermons, prayers of praise, prayers of petition—but not much listening. When I encountered Buddhism through college reading, I thought, "Oh, here's the listening piece." That was in the late 60s or early 70s. Today many congregations incorporate some form of meditation; back then, meditation was not mainstream. When I was drafted, my Buddhist practice informed my beliefs, enabling me to obtain conscientious objector status, becoming an Army medic reflecting the principle of ahimsa. But without a community, the meditation practice I developed with the support of a Japanese Zen monk faded out after five years.

Around 2008, as my children were older and my professional time allowed an opening, I returned to meditating. My professional life (the increased environmental emphasis) and faith life have grown together and inform each other. Together they strengthen the sense that we live on a beautiful but fragile planet and must do something to save it for future generations.

We're all part of the same planet; what can each of us do? There is a sense of participation and sharing. Responsibility and love grow from this awareness. If we don't participate in a movement toward wholeness and unity with creation, some dark aspects of human nature will dominate. We see in the news what we get if we let that happen.

Stephen Jay Gould, a Harvard biologist, observed that we homo sapiens are not the top of the tree; we're not the crown of evolution. Life on Earth is more like a bush; we're just one branch of this bush. I'd like our branch to expand, but I don't believe its continuation is preordained.

I'm agnostic. I don't believe we're the most favored species. The planet is only at half its eight billion year expected life, and well more than halfway beyond what is expected to be its habitable life. What will happen in the next billion years is unknown. I hope that our species can continue because we have so much potential for good—and yet, we have so much to overcome at the same time.

I was out on a walk the morning of our discussion along an old section of California Highway 1 roadway known as Devil's Slide. I walked above the shore through a stretch with visible sedimentary layers as well as granite. It reminds you of what this planet has been through. You can see granitic rock, and realize this was upwelling magma from within the Earth as tectonic plates slid over one another. Then you see the sedimentary layers all buckled and folded, laid down by layers of ocean and beaches, and ocean and beaches over millennia. One begins to consider the flow of time. We are here for such a brief period. I want our species to do well, but even if we don't, the planet will survive, and life will survive.

For Further Reflection

- Would you consider switching careers because your job did not match your values?
- Gerald had several unexpected opportunities to pursue new venues for carbon reduction. Do you share his willingness to "move with the flow" as surprises come your way?
- Do you associate with people of different faiths? What are the advantages, and what have you learned?
- Where do you see the link between your faith and your work?
- Do you agree with Stephen J. Gould that we are not at the top of the evolutionary tree? If so, what conclusions do you draw from that?
- Do you find it reassuring that our planet will survive, even though humankind may not?

Bill Myers

When I joined the Peace Corps in the 1960s, I was assigned to the tropics to work as a volunteer with indigenous people in the Brazilian Amazon. That was my first exposure to a very different lifestyle from what I was used to in the US.

After I left the Peace Corps, I started working for the state of California on agricultural labor, developing policies for farm labor in California. However, in the early 1990s, I was asked to lead the Peace Corps involvement in the Amazon and Brazil—selecting volunteers, developing programs, and overseeing contacts with native people, scientists, and visitors. Over time, it became clear to me that life in the US was not a good model and harmed indigenous people. The way we built cities, organized our daily routines, and related to each other, everything we called "civilization," didn't fit. If we imposed that life, people wouldn't flourish.

In 1992, I was asked to represent UNICEF at the Earth Summit in Rio de Janeiro where climate change was formally recognized by the UN. There had been prior meetings at an organizational level, and of course climate change had been recognized internationally since the 1880s. But in 1992, the UN decided they needed to take action. They called it the "Framework Convention" because they laid down a framework for an agreement to deal with climate.

From then on, I began to see climate change as a policy issue. I had been involved with jungles for a long time and knew that bad things were happening, especially regarding violence against children. Climate change kills children. They become malnourished. They can't get the food that they need anymore. They die early from pollution and diseases. The fossil fuel industry kills between 5 and 10% of the world's population every year. Of all the people who die, about half are children, most of them under five. It's overwhelming. Climate change affects everybody, of course, including old people, but it disproportionately affects children—and, of course, generations to come.

Belem is the main port city for the Amazon. The COP30 conference is going to take place there this year. I lived in that city twice: from 1968 to 1970 and again from 1990 to 1992. The second time, I helped develop a project. People had the idea that we could go in there, take down the jungle, and plow the fields, just like we did in Europe and the United States. We would grow more food to help the local population. However, the problem is that tropical soils don't act like temperate soil. A Catholic Church organization had started a project on one of the islands. They had taken down all the palm trees, gone in with tractors to plow up the land, and ruined the soil. Not even the palms would grow again. The Westerners thought they were doing people good because they were showing these primitive people how to live like Europeans.

I had made that same mistake earlier, in my Peace Corps days, trying to show the locals how to do agriculture. We were going to make these really poor people better off by having them use our methods for productive farming. Of course, the soil wouldn't carry what they planted, and now the indigenous people no longer had their coconuts.

So in 1992, I worked on a project to reestablish the jungle so it would again produce food for the locals. Agriculturalists who had been raised in the jungle came in and began to take vines out and cover the ground to give it shade. They started to reestablish that area of jungle, putting in plants that get along and produce food to keep people from starving.

In Belen, you walk to the edge of the city and all of a sudden you're in the jungle. It's not like our cities that go on forever. They have a big museum visited by all kinds of people: Europeans, Americans, indigenous Brazilians, botanists, etymologists, researchers, and tourists. They all go there to study the jungle. I got to know many of them because I lived there and saw them in bars or in town. They were absolutely fascinating.

They had a special project for what they considered to be a really flourishing piece of virgin jungle. They had been studying how the species related to each other and had done all kinds of botany research, working on a 30-hectare tract where there had been a forest fire many years ago. They decided they would leave that tract alone to observe how the jungle grew back without human interference. So they divided the area up into one-meter squares, and every year or two they counted everything and entered the data in their books like true scientists.

While I was there, they realized that, surprisingly, the 30-hectare tract wasn't going to look anything like the jungle around it. It was going an entirely different direction. They had no explanation for that. They gathered around a big conference table in the museum to discuss the matter when somebody raised the question, "Do you suppose what we have been taking for virgin jungle is not virgin jungle? Could it have been cultivated by Indians long ago?" The minute he said that, everybody knew that was right. It just had never occurred to us because Indians don't do agriculture like we do in fields. They carefully observe how plants interrelate. They cultivate the relationships between everything—plants, animals, land, and sea. The scientists saw a route to follow: Where the fruit had fallen into the river, the fish had come to eat the fruit, and the people had caught the fish. What the scientists had taken to be virgin jungle had been beautifully cultivated over the years to produce food and take care of everybody while at the same time protecting the jungle.

We "civilized people" had been disrupting ecosystems by putting fields out there to grow food we were used to.

We were drying up the rain. We were actually affecting the climate itself, determining when and how much it rained. When I returned to Belen 20 years after my first time, it was not the same climate, and native people were starving. Suddenly I realized how smart the indigenous population had been. They had developed a sustainable form of agroforestry, preserving trees and climate and a natural environment that provided them bountifully with a rich diet.

People have forgotten the old ways. We are out there now, cutting the jungle down and putting in fields for cattle. The American and European agricultural system is changing the climate and it's killing the jungle. The changes are dramatic in the tropics. I saw the impact on children, and I helped design a project to reestablish the environment so that people can grow their food again.

To me, this is not theoretical. It's a very visceral issue that affects real people and their livelihood.

One more thing about that. When I was a Peace Corps volunteer in Panama, I was assigned to an indigenous area where Indians lived not on the land that had all the mosquitoes but on a series of low-lying islands right off the mouths of rivers. The rivers came out of the mountains and went out to sea. Just a quarter mile off the land there were these sandy islands. The people who lived on them were very healthy, almost completely free of disease. I visited them in the villages on these islands. But now they have to leave them because the sea is rising. They're among the first people in the world who have to leave their homes in order not to be flooded out. Climate change is destroying people's lives. I saw it in the Amazon and in Panama. It's not a theoretical thing for me.

I grew up in a culture that came out of the Church of the Brethren, one of the three peace churches along with the Quakers and Mennonites. When I was fifteen, I decided I would not have a regular career but do only things that served peaceful purposes. I was already aware of and impressed by the violence in our society. When the Vietnam War came along, I became a conscientious objector and ended up in the Peace Corps.

The Brethren are mostly back East, so I am now a member of a Methodist congregation. The Methodist denomination is rich. They have something like 28 billion dollars invested in fossil fuels. From my faith perspective, that didn't make sense. I met with a pastor who had become very interested in climate change and got to know her. When I joined the Methodists, I brought a piece of Brethren culture with me into the church, helping them become a little more activist and press a little harder. My pacifism stems from my Church of the Brethren background.

Our faith says that we need to be guardians of the planet and that we need to love one another. I make a distinction between following the rules and doing right because of active love. Sharon Delgado wrote a book some years ago called *Love in the Time of Climate Change* where she laid out all the good reasons why we should get away from fossil fuels. It's beautifully written in terms of what the rules are, but it doesn't say why we should obey them.

Take a look at the Good Samaritan story. The reason why the Levite and the priest passed by on the other side of the road rather than help the wounded man was that they were following the rules of religion. The policies said you shouldn't touch an unclean person. The parable makes a distinction between following the rules and paying attention to a person out of love. Religion in this case was getting in the way of loving. Perhaps rules are a kind of armor to keep us from feeling the pain of real people, real children. I have seen children in the Amazon who are suffering. I care about that. But people who go to my church here don't see that. They're not next to it.

I am an elder now. In traditional societies, until recently children and youth had to get all their information and values from people older than them. You had a high-school teacher, and in college you had to learn from your professor what books to read. That's not so true anymore. We have easy access to information. Any high-school student can get the same information that older people can find. They don't have to go through adults for that purpose.

I am critical of the idea that aging and saging go hand in hand. That used to be true, and it continues to be true for certain things that require life-long experience, like dealing with difficult family relations or settling disputes. There are things you learn only as you get older. But that's not true for science information. Any teenager can learn as fast as I can, if not faster.

When it comes to climate change, there is no evidence that older people are wiser or know how to handle it. We get interested in technology, for example, thinking that technology can make the world better. Well, that may not be very smart. Many social scientists and a lot of young people say that the problem is our values. As long as we let technology determine our values, we're going to be in big trouble. We need to do a lot more than that.

Young people today go to conferences where they talk more about values and what kind of society we are. They don't necessarily come out at the same place that people our age do. There's room for both, of course. I'm not saying one's right and one's wrong, but they're not the same. So what can be done? Well, we can share information. We can get together and talk about it.

It may be hard to have those conversations in the United States. Our culture gets in the way of elders and youth talking together to share information. Other places have an easier time with this.

We need to change the model of being an elder, moving away from "I'm going to show you" or "I have the power, so you do it my way." We have to trade that role for another model that many cultures have but ours doesn't. It's the role of the elder as a weaver who can establish relationships across all the different levels of society.

In Africa, children would never contradict or openly disobey their parents. If they have a problem, they go tell a grandparent. Then the grandparent steps in and talks to the parent. The grandparent is responsible for weaving a good relationship between the generations. That's an alternate way to look at elders. They are weavers, not power brokers who lay down the law to live by, but working with different parts to solve problems and maintain relationships among generations, children included. Woven together, they all contribute.

I have lived and worked with indigenous groups, and I've actually seen that process. It's a skill we would have to develop. We Americans are good at the extreme opposite. How do we share information without exercising our "power over," just pushing legislation and rules? How can we use diplomacy, weaving together a society that makes room for kids and their ideas, proceeding in a way that makes everybody a participant?

We all have to work together to solve the problems from climate change. We don't need to argue or get defensive about it. Elders are exactly the right people to promote that way of dealing with conflict.

For many years, I was on the board of an organization called the "International Institute for Child Rights and Development," which is centered in Canada. Four times a year I had to go to Victoria, British Columbia, to attend board meetings. The person representing the indigenous peoples was a guy named Bill White. He told me, "I applied to be an elder and the elders turned me down. They said I have too much warrior left in me. You're a great warrior, and we need warriors to defend our rights and push our causes. Do that for now, and we'll tell you when the time is right for you to become an elder." So he continued that work specifically for them, and because climate change is a very big deal, he later asked how they could work on climate change together. Then they made him an elder.

There are some things that we can learn from tradition, like that alternate way of being an elder; and there are other things we can learn from innovation, like ideas the youth can bring into the conversation. Bringing multiple generations together makes us wiser as we grapple for solutions. For Americans, such collaboration can be difficult because we're very individualistic. We say, "I can take care of myself. Thank you." But that's a lie. That's a delusion, and in this world more than ever.

We have to change who we are and what we value. I don't think we can save the planet just by developing a fancier version of the good old capitalist industrial society, living in big cities and seeking to accumulate wealth. We need to rethink what we understand by progress. What should we value? Can we measure worth by who we are rather than by what we have? We need to realize that we're part of everything and everybody. Once we do, what behaviors help us carry out those values? So far, we have gone for the wrong values, as if life is all about "stuff" I can get. That's wrong, and if we continue, we'll kill ourselves and our children.

The most important lesson I ever learned as an adult was on a trip I took with an African. I was in Nigeria to look at the problem of kids working on the street. Were they being violated? Did they get any help? The person I was assigned to travel with in the jeep was a famous Nigerian sociologist with a Harvard degree. When I asked her what she thought about Harvard, she said, "Oh, I liked Harvard a lot, but I was puzzled by Americans. Americans don't know who they are."

I thought about that for a minute, then asked for an explanation. I pestered her until she stood up, put her long bony arms up over her head, and said, "Bill, I am born within a certain culture. I speak a specific language. All of this is part of who I am. It comes from my ancestors." Then she put her hands out sideways. "How would I have the food I eat or the house I live in? Who would stand by me when I am sick? What would I do if I didn't have all these people out here providing for me? They are part of who I am." Then she lowered her hands down between her knees and said, "I'm the ancestor for my children and my grandchildren and generations after them. I have to watch how I speak and what I do and how I deal with them. They're part of me, even those who aren't born yet." And she looked at me and said, "Bill, there is no 'me.' There is only a 'we.'" That conversation completely changed the way I see the world.

For Further Reflection

- What did Bill learn through his work with indigenous people?
- Which part of our "civilization" is most harmful to the global community?
- Do you know examples from North America where white immigrants unwittingly destroyed habitats?
- How did native agriculturalists in Brazil reestablish soil health?
- Do you agree that America's problems are linked to individualism?
- What role does Bill advocate for elders? How would it work?
- Do you like the lesson from Bill's African friend? Give an example of moving from "me" to "we."

Cliff Kindy

In the 1980s I was asked to be a Church of the Brethren representative on the steering committee for Christian Peacemaker Teams (CPT). The program is now called "Community Peacemaker Teams" because people from different faith traditions are also invited to participate. I joined before CPT had sent anybody out. We first looked at what might happen when we send people into conflict zones and at how we would train people.

In 1993, Gene Stoltzfus, the director of CPT at that time, came to me and said, "Cliff, we think you should join the team that's going to the Gaza Strip." I quickly said, "No, it's not possible. We have two young children. I can't go. Also, it's the busiest three months for our gardening work. That's how we make our income. Thank you for thinking of us, but no, we can't do it."

But Gene was persistent. He came back twice again and said, "We really think you should be on this team, Cliff." At that point Miriam, our younger daughter, said, "Daddy, maybe we should make it possible for you to go. Many fathers and sons leave home in Gaza and never return. You are just going for three months." So Daddy went.

The experience in Gaza then was much different from what it is today but still brutal. I went and it changed me. It gave me faith in the fact that, if we're willing to put ourselves on the line, if we're willing to offer whatever we have, even if we don't have all the skills, languages, techniques, and experience with nonviolence, God will use that offering, that tiny offering to work a miracle. It doesn't depend just on our expertise to bring nonviolence into a violent setting. That was the beginning for me delving into deeper work with CPT.

Arlene and I started with tax resistance. The reason we ended up trying to limit our income was that we were at a disagreement about whether we should just not pay the federal income taxes or whether it would be better to lower our income so we didn't owe taxes to begin with.

I had come out of Brethren Volunteer Service. During the Vietnam War, our team had decided not to pay the phone tax, because most of that money was going toward the war in Vietnam, according to Chairman Wilbur Mills of the House Ways and Means Committee at that time. So Arlene and I became tax resistors by keeping our income below the taxable level, and eventually I became a draft resistor as well. I had registered as a conscientious objector, but then after a year I returned my draft card saying I couldn't even participate with the system of selective service because its primary motive was to collect bodies to go to Vietnam. I didn't want to support that at all.

CPT was a nice alternative. I was going to war—not to expand or continue the war, but to stop the war. Experiences in places like Vieques, Puerto Rico, or Chiapas, Mexico, were dramatic examples of nonviolence. ssisting local partners, we turned swords into plowshares. We confronted the war system and turned the space into a place of justice and peace.

In Chiapas, the community was 90% indigenous. 70% of the people had been displaced from their homes by military and paramilitary operations. In four years, we were able to watch as they returned to their homes. The indigenous Las Abejas Group, Mayan indigenous people, trained their own barefoot doctors and their own teachers. The Mexican government wasn't providing that kind of care. The Abejas learned how to develop a product that had a market while in the rest of the state, hundreds of thousands of people were fleeing to the US for jobs. They couldn't support themselves after NAFTA had crashed the corn market and the coffee market in their communities. But the Abejas figured out how to raise organic coffee and shade-grown coffee to maintain their markets so that they could stay with their families. When we left, most of them had returned to their home communities after four years.

Vieques, a little island east of Puerto Rico, was a training ground for most of the troops that went to Vietnam and also for the troops that went to Iraq. And it was a bombing zone where the military could test the practices of war and where munitions manufacturers could test their tools on a piece of land. But people also lived on Vieques, sandwiched between the bombing zone and the munitions stockpiles on the western end of the island. The people were developing high rates of cancer and leukemia. Their babies had birth defects. The fishers no longer could find fish that weren't poisoned by the toxins from the chemical munitions used in the training.

So a campaign started. CPT wasn't the main promoter of the campaign, but we became a regular part of the protests as we took nine delegations to Vieques over the next years. The dramatic part was that our resistance community set up tents as a way to stop the bombing. Whenever we set up tents in the bombing zone, the Navy would stop bombing. Next they would come in and try to remove us or arrest everybody, and then they'd go back to bombing.

The admiral of the Navy said, "We'll never leave Vieques. There's no place like this anywhere in the world." But we kept setting up tents in the bombing zone. And getting arrested. There were huge demonstrations on the island itself. The resistance campaign was joined by Catholics and Protestants, women's groups and youth groups, veterans, fishermen, and independent political parties. It was an amazing movement. And even though the admiral of the US Navy said, "We'll never leave Vieques," after three years they pulled out of Vieques and also closed the largest naval base outside the continental United States, Roosevelt Roads on Puerto Rico.

I keep saying, "You know, we have in our hands much more powerful tools than the tools of war. Look how the most powerful military force in the world crumbled before a few people who set up tents in the bombing zone."

So my main focus has been on dismantling unjust structures, undermining and counteracting systems of domination, and protesting against any form of violence. That's where my passion lies. Of course there is also a link between violence and climate. War zones are among the worst climate disasters. Think of the resources that are spent on the destruction—the toxicity that remains on the ground, in the water, and in the air after a war. Wars are huge stumbling blocks to maintaining a world and a climate that is healthy, safe, and sustainable.

Wars provide the financial impetus for corporations to keep operating. They sell to anybody and everybody without limits, without restrictions. But wars don't build anything up; they don't help anything survive; they don't bring goodness to society. Wars cannot be in any way constructive.

In February 2025 we campaigned against Raytheon in Fort Wayne. Raytheon is the second largest military contractor in the world. They provided the weapons that destroyed Gaza. Interestingly, they also provide weapons to China, Russia, Iran, and Lebanon. In our attempt to stop the weapons from flowing out of Raytheon, we tried stopping the war effort on both sides.

We organized a big event to witness for truth and peace. Christians, Jews, Muslims, people without religious faith, old people and youth, veterans and peacemakers—all gathered at the Parkview Raytheon plant. There are eight military contractors in just the town of Fort Wayne. We took written invitations to those eight military contractors and to the Air National Guard Base in Fort Wayne, all complicit in what's happening in Gaza. We presented three requests:

1. Stop selling weapons or technologies to be used against the people of Gaza so that the genocide can end.
2. Use your influence, which is greater than our influence, to encourage the US administration to implement a total arms embargo on Israel.
3. Figure out ways to convert your production lines from weapons that kill the people of the world to tools that help the people of the world.

Those were our three simple asks. And when we didn't get any response, we tried something else. I have a friend, Mike Ferner, who served as a Navy corpsman in Vietnam and later became the director of Veterans for Peace. He was also with me in Iraq when we were trying to stop that war before it started. He asked me, "Cliff, I know you're not a veteran, but do you want to gather your friends and be part of this campaign to try to end the genocide against the people of Gaza?"

His group sent an eleven-page letter to the US State Department, asking them to halt weapon shipments shipping weapons to Israel because were breaking five to seven different U.S. laws by promoting that genocide. Nobody responded. The group contacted 17 State Department offices across the country and received no response.

He said, "We're going to go into our local communities, to congressional offices, and to military bases to continue this campaign. Will you join me?"

I said sure, because I'd been trying all kinds of things. We'd been to congressional offices praying and singing, written letters to the editor, and called the White House. We tried everything. At one point we strategized how we could combine the efforts of the Vatican, Muslim Peacemaker Teams, Rabbis for Human Rights, with Gulf State funding and jointly with Brethren Disaster Ministries. We thought Children's Disaster Services teams placed on the ground in Gaza might be the beginning of a successful campaign against the genocide.

I got rejections. So I said, "What else can we do? What can we possibly do?" And then we tried to stop the weapons symbolically. We got arrested and asked for a jury trial. Then the prosecutors dropped the charges. The judge had to sign off on that decision. Usually that happens within a day or two, but in this case, the judge finally signed off only a week or two later.

We said, "Well, since we can't be visible at a trial, we'll call a press conference." We had the media and lots of other people present. Now we have a YouTube video of the press conference that we're sending out to lots of different places. And it did get picked up by the media.

Years ago, in 1997 or 98, I tried a different kind of protest, both in Israel and Iraq. In Israel-Palestine, we started a hunger strike as a CPT team when we discovered that 700 homes were scheduled for demolition in the Hebron district. The policy was to make it impossible for Palestinians to live in the West Bank, so that they would move out. We felt helpless. Apparently, we were not able to make a difference. I mean, we had already been arrested for chaining our bodies to the houses. We had tried to rebuild houses. We had tried to stand with families who had home demolitions scheduled. Now, in a last attempt, we tried the hunger strike.

On the last day of the strike, we got arrested. A Jewish Rabbi and I and two Palestinians were arrested while rebuilding a home that had been demolished a year earlier. I was kicked out of the country, taken to trial on Easter Sunday, and taken in leg irons and shackles to the airport to be shipped home.

From that point on, there were no home demolitions for a year and a half. Zero, at least in the Hebron district, maybe in the West Bank. Evidently some good happened because of that strike.

And then the Iraq War happened. We were in Baghdad when the "Shock and Awe" bombing came out of the sky. It was sobering. We were witnesses on the ground. Much of the international media had fled Baghdad because of the danger.

After that bombing, the US occupation started in Iraq. Thousands of homes were invaded in the middle of the night. The military practice was to have every male in the house on the floor on the ground, blindfolded, hands tied behind the back, within three minutes or so. Lots of people were detained, sometimes whole communities. Every male from 15 years of age to the 90s was detained in one village, and almost always with no charges and no access to a lawyer.

So we began a fast to make that practice visible, which was turning every US person into an enemy of the people of Iraq. Home invasions were universal across the country, in the rural areas, in the villages, in the cities. Tens of thousands of men were held in detention. So we wanted to make that visible and eventually worked with Seymour Hersh, the investigative reporter. That brought much more attention to our documentation efforts. We met with some of the highest-ranking US officials, both military and civilian. We explained, "If you continue this detention process, you're going to lose any friends you might have made by getting rid of Saddam." At first the people of Iraq who talked with us had said, "We're glad that Saddam is gone." But then, after home demolitions and administrative detentions and the torture that was happening all across the country, they said, "You know, what we had under Saddam was better than what we now have under Bush." At that point, anybody who was connected to the US became a target, an enemy.

For that reason we moved to northern Iraq instead of living in Baghdad. And that's where we began our Lenten Fast, liquids only. It's a spiritual discipline. I think for nonviolent action it's important to have spiritual grounding. For me, fasting, or the kind of pilgrimages I did when I walked across Ohio and Illinois, are important pieces of staying connected to the spirit in a way that keeps us alive and powerful.

I have all along worked with people from other faith communities. Christian is my tradition, but others are on the same path: Jews, like Rabbis for Human Rights and Peace Now groups in Israel. Muslim groups in Palestine, Gaza, or Iraq. Indigenous faith groups in Chiapas. That's part of the reason we at CPT call ourselves "community" rather than "Christian."

When I think about taking action on climate change or other justice issues, I see it as an invitation to goodness. And I recognize that the invitation has to include myself, but it also includes the "enemy" or the one we perceive as "other." Raytheon. Big Oil. Boko Haram. Folks in the military or occupation forces. Presidents or prime ministers. My experience has been that those of us who appear to be on different sides of an issue can genuinely come together. We can come to a place that raises us to a higher plateau that neither of us alone would have been able to reach. We come from a different set of experiences, but if we can come together genuinely searching for goodness or for peace or justice, we find we can reach a level we never would have reached before, a different understanding of our world.

I approach others with that kind of understanding. It's not an attempt to destroy the other but to interact with the other in such a way that the other can maintain her or his integrity, ingenuity, and dignity. We can find something rich, powerful and just, peaceful and healthy for our world.

I remember years ago here in North Manchester we were working on what became Education for Conflict Resolution. We were active in the schools and communities around us and elsewhere across the state. I was working with a veteran who had received Purple Hearts for his service in Vietnam. He was a staunch veteran who often would step in when returned soldiers lost it and holed up in a house or a woods somewhere. He would go in, talk to them, and help them through post-traumatic stress. We invited him to be part of our core committee. He and I spent lots of time together on my front porch, or we drove around together. The two of us were the ones who brainstormed and came up with the idea of doing something in the schools with the children. That was the beginning of Education for Conflict Resolution. So that was a powerful development.

Another time I was asked to go to Nigeria to represent Brethren Disaster Ministries. Ekklesiyar Yan'uwa a Nigeria (EYN), which translates to "Church of the Brethren in Nigeria," is a denomination that emphasizes peace, community service, and the teachings of Jesus. EYN was facing a situation where Boko Haram slaughtered Christians. The denomination had lost a thousand church buildings. 10,000 EYN members had been killed. 600,000 had been displaced by the violence. Boko Haram controlled the same areas where EYN had the most members.

Before I left for Nigeria, I learned that Boko Haram had kidnapped 300 young women in Nigeria who were attending school. I started a group of US adults who were willing to be exchanged for the release of the captives. They would go to Boko Haram and say, “We are ready to be taken hostage by you if you let these others go free.” The proposal was rejected by the Church of the Brethren Annual Conference after somebody argued, “Oh, but if you do that, it'll probably interfere with what the US government is trying to do to resolve the issue.”

I imagine that had we been able to go to Boko Haram, they would have asked, “What in the world inspired you to be willing to be exchanged? What kind of faith do you have? Where does your faith come from?” And then we could have talked about that. We could have asked them in turn, “Well, what inspires you and how do you think goodness is going to happen in this space?”

If we had talked together, think of what could have happened instead of what did. Here we are, 10 or 11 years later, and the same violence persists in Nigeria.

When we work on a cause for justice, it's easy to forget to talk to the other side. We have to shut down this pipeline. We have to save this sacred space for the Apache. The copper company is the enemy because they want to dig a trillion dollars worth of copper out of the land underneath a sacred space.

So how do we come together? Maybe we need to include in our conversations what we can do so that there's a world here for our grandchildren, so that this ground can grow food to feed people, so that this water will be safe to drink. Maybe others have expertise way beyond what I have, but together, we can use that expertise or other people's experiences to find a space that has goodness in it. I think that coming together is important. It may include resistance, like parking the tents in the bombing zones, or parking the tents in the path of the pipeline, or accompanying the Apache stronghold and that sacred space that's about to disappear in a hole in the ground. Coming together in our efforts is what gives me hope. It inspires me to keep going, because it has worked so frequently. War never does.

I'll say one more thing. I'm working on it for a sermon. It's also the title of my book, *Resurrection Peacemaking*. In the Christian tradition, we remember Jesus's resurrection as a defeat of the powers of sin and death. The empire did its mightiest to win. It nailed this rabble-rouser to a cross until he was dead, then had him thrown in a tomb, rolled a stone over the entrance, and sealed it with the signet of the emperor. Then guards were put there to make sure nothing untoward would happen. But something untoward did happen. The power of non-violence is always about love, about goodness, about light.

Empire tells us, "Don't speak about the suffering in Gaza, don't interfere, don't mention what crimes happened in Iraq, don't try to close down the Vieques training ground. Remember that we hold all the power. We can send in 5,000 troops from Fort Bragg. We can arrest everybody. We can control the news."

Well, maybe it can't control all the news. And that's where resurrection happens, when some things start to reappear, right out of the solid concrete ground that the empire represents. And suddenly the picture changes. The seed that is covered up by the concrete can start to grow through the crack, out of the ground.

I see on our plot here, at Joyfield Farm, how the seeds are growing up through the crack in the concrete. The chemical wasteland that sprouts corn and soybeans all around us leads to a dead end for the soil. But near my home, the soil begins to show new life. The birds come back, insects and frogs live here, and butterflies return. They remind me of resurrection.

For Further Reflection

- Many elders now involved in environmental advocacy started as anti-war protesters in the 1960s, 70s, and 80s. What do you see as the connection?
- Which of Cliff's actions do you find most inspiring?
- Describe the methods used in Vieques and Chiapas. What made them successful?
- What does it mean to "stand with the oppressed" in today's context?
- Review Cliff's "invitation to goodness" as he talks to adversaries. What does he recommend for shifting the playing field from enmity to partnership?

- Cliff is a radical peacemaker, firmly rooted in the tradition of Gandhi and Martin Luther King Jr. Do you know people like him? Share their story.

A Wish for What is to Come

For Future Generations

My child, I want a world for you
we have not had before,
where people build community
and no one thinks of war,
where we preserve the Earth for future generations.

My child, I see a world that's just,
where all can dwell secure,
a world where need and hunger cease
and people help the poor,
where we will stand together with future generations.

My child, I wish you could grow up
untouched by greed and hate
within a world community
that's whole and unafraid
since we protected Earth for future generations.

My child, I want to rest assured
that after I am gone,
more generations fill the Earth
and life continues on
with guardians of the Earth for future generations.

Part 7

A Matter of Faith

Voices:

Gregory Stevens
Jan Warren
William Morris
Eileen Altmann
David Howell

Our country is in a dangerous period of transition. Many institutions that we trusted to be indestructible are being undermined or shattered. A big rift divides our society, leaving people insecure and contributing to a drift toward authoritarian leadership.

How do we get out of this morass? Will we voice our opposition against self-centeredness, violence, fraud, racism, and injustices of any kind? Will we see it as our moral and spiritual responsibility to be courageous and bold, engaged to make the world a better place by working for equality, safety, and compassion for all?

Most spiritual traditions have a key component that teaches love of neighbor. This love does not just mean being friendly. It can require risky action in the face of wrongdoing. In the Hebrew scriptures, prophets remind people to do justice, be merciful, and walk humbly with God. Christians claim Jesus as the model of how to live, not just by healing, feeding the hungry, and siding with the poor, but also by confronting those who cause oppression of any kind. Jesus was speaking up for the underdog—the imprisoned, the sick, the outcast, the immigrant, and the stranger. He pointed out neglect and gave dignity to the lost.

Can we regain a prophetic spirituality that challenges those who abuse power? Our world waits for committed people everywhere who refuse to participate in destroying the environment, in abandoning a global community for the sake of national self-centeredness. We must say no to the pollution of the skies and the poisoning of the oceans. We must speak up against the mistreatment of women, immigrants, and people laboring for unfair wages. We must stop destroying our Earth for the sake of more wealth and power in the hands of the few.

Prophets speak the truth to a culture of lies. At the same time, they say yes to equal rights for all. They work to alleviate suffering. They spread a message about unity and care for one another. Our moral and spiritual responsibility is to help each other get out of the current morass. We can begin by speaking words of truth in public. Civil discourse—not one that spews hatred and division—is an effective tool for change. All we need to do is calmly state what we stand for, not to win an argument, but to expand hearts and minds. We can talk with each other about what we understand to be beautiful, kind, and loving. We can listen respectfully to a person's spiritual idea and learn where it came from.

As moral agents, we make decisions. We can choose the selfish way, saying, "Not me. I'm not getting involved in that. Let others handle it. That's not my problem." We can choose passivity, stating with a shrug of our shoulders, "I'm sure it'll all come out all right." The third choice is to engage in political and spiritual action. We form or join a group to deal with an issue that requires change. We study it, develop ideas, and support the cause by saying, "I want to take part in this effort for transformation."

What if we spent time each week in conversation with friends about what we are on this Earth for? What if we asked ourselves what we could do next in order to show kinship with all living things? And what if we promised to stand with each other as we begin this journey? In this transition moment, we need to bring our concern for the Earth into the public conversation. Each of us has a part to play. Each of us can listen for prophetic inspiration, find it within ourselves, and step up to show uncommon courage.

Martin Luther King Jr. said, "Faith is taking the first step even when you don't see the whole staircase." In this chapter, we hear from people who realize that our planet is in danger and who show how their faith has led them to take decisive action.

Gregory Stevens

I grew up as a Baptist evangelical Christian who wanted to be a minister. In college, I started to learn about climate change and climate justice, an all-encompassing subject that was interesting to me because it connected to almost every other justice issue. At 18 or 19, I hosted a conference and had speakers fly in from all around the country—which might not be the best ecological thing in retrospect—but anyway, they flew in and we had this great conference, where I met Tripp Fuller, whom I had known through podcasts and Twitter. He introduced me to process theology. I was fascinated by the problem of evil at the time, and process theology has an interesting answer to that. So I packed my bags and flew to California for grad school.

That's when I amped up my climate activism. One of the professors at Claremont, John Cobb, wrote one of the first theological texts on climate change and religion, called *Too Late.* He wrote it in the 1970s, already asking whether it was too late! I attended seminary from 2007 to 2012. And so all those years later, I'm sitting there with this dynamite process theologian Tripp Fuller next to me, trying to ask Jeffrey Kwan, the president at the time, to divest from fossil fuels. The school had some bad financial problems, so they weren't quick to do that. But it was a start to my activism, trying to organize other students, getting petition signatures, and having hard conversations with people.

By the time I was done with seminary, I didn't feel like I was a Christian as much as I was interested in faith and religions. I took a job at a church although I felt too young. Being a pastor should be more like being a wise person, a sage, someone who's lived a little. At my church, I want someone a bit older than me to give advice. So I left religious leadership work and went into union organizing, which was helpful for boosting my activist skills. I was trained in theology, but I didn't have the activist skills that I needed: learning to make phone calls, organizing conversations, pivoting at the right time, and being in the field for difficult discussions.

But even though I didn't want to be a religious leader or a pastor, I missed being in the religious world. Then a job popped up with California Interfaith Power and Light (CIPL) for a religious person interested in interfaith work in ecology. This was a great opportunity because I didn't have to be the minister but could still work with religious communities on issues of climate change. I could blend organizing and union labor activism skills with my theological grounding and interest in other religious traditions, and all that without anyone coming up to me at the end of a worship service saying, "How do I raise my kids?" The question would be more like, "Where's the petition?"

I have found my job to be a great blend of both activism and faith. You know, religion gets a bad rap lots of times. I get to see the best versions of it: Jews, Sikhs, Baha'is, and Christians putting solar panels on their places of worship; congregations planting trees; having ceremonies to reflect and take note of the ways we've harmed the natural world, how we are the natural world, how taking care of the Earth is taking care of ourselves.

Another thing I love about interfaith work is that I always learn more about other faith traditions. There are certain things you just don't know. Here's an example: I was invited to give a talk at a Jewish synagogue, which was a big deal for me, and the board chair of the synagogue nudges me and says, "Get off your phone." I'd been reading my notes on my phone because I was about to give my little presentation. I know that it's not always kosher to have your phone out in public, so I figured he meant, "We're at worship, put your phone away." No, what he really meant was "We're Jewish and we don't use technology on Saturdays. It's Shabbat." Well, I didn't take the clue. And so I get up there and I give my talk with my phone in front of everyone in the most sacred spot. And the Rabbi comes up to me afterward and says, "You know, that was really great, but maybe next time bring written notes because we try not to use technology."

It's fun things like that, when you think you know your Jewish friends and suddenly become aware of a cultural difference. It's a bit of a shock, but it shakes your reality. We are both different and similar. You know, when I'm in the synagogue, it feels a lot like church. We're standing, we're singing, we're in community together. But little things are easy to miss. And so it's a great learning experience to go into a Sikh gurdwara or a Hindu temple and realize that you bowed at the wrong time or said the wrong thing. I have so much to learn, and this world is so rich in diversity. How beautiful and magical is that moment when you realize you've got it wrong but opened a door to learn more!

Now I'm a Unitarian Universalist. One of our values as a denomination is realizing the interconnectedness of all things, which demands of us to be in relationship. It's saying that because we are interconnected, we're our neighbor's keeper. It's putting in sacredness. It's making it holy. Everything's in process and in flow together, as the old butterfly wing analogy states: What you do over here, no matter how small, is going to make a big impact somewhere else.

Fossil fuels are not something you see every day with your eyes, but their use is destroying the world. When we take note of interconnectedness, we start to see how what we do to ourselves or what we do to others is going to affect all of us. There's a sacredness to protecting, caring, and nurturing all of us rather than harming and creating division.

When we see ourselves as interconnected, as a part of this larger system, we're far more likely to take action. We are less likely to seek domination or to destroy life.

Abrahamic religious traditions have certain narratives implying, "It's your land, you control it, you dominate it." In contrast, more indigenous and native traditions understand that we're a part of this natural world. We should revere the sacredness of being interconnected with the land and each other. We shouldn't pretend to be more special or more unique than others. Once we adopt the indigenous view, we start to act differently. We vote differently. We see policy differently. We see pollution differently. We see the animals running around in the wildfires, and we see the people suffering from the devastation, and we see that Altadena was an all-black neighborhood that burned out. We notice all of these connections. When we look at what is happening in a more holistic way, our responses are going to be healthier.

Sometimes if you just poke at one little thing, you don't see the larger context. You might be treating the symptoms but not the actual core problem. When we realize the larger social economic structure, we can get to the core: an economy that's based on the endless extraction of resources on a finite planet. Once we recognize that larger system, our responses might be different. Reusable water bottles are great, but that's not going to stop the system that's creating endless water bottles.

Religious diversity is helpful if we act with inclusion and respect, especially when we're coming from different places. One of my favorite analogies is to think of religions more as languages. You would never say Spanish (or some other language) is wrong. You would just accept that it's different. It comes from a different part of the world, and it makes so much sense that it looks and smells and tastes like that part of the world. You know, it is actually really fun to be around people with different languages, and you say bird and then everyone has a word with a similar meaning, but it's not a direct correlation. All of a sudden it's not just a bird, it's this cool figure that you've imagined in your head.

Religion is like that. When you get a bunch of religious people together, if you start talking about peace, a Sikh may think back to insane marginalization and oppression, and Jewish folks remember persecution. If you bring them together with an attitude of respect and inclusion, all of a sudden the barriers break down. You no longer start with differences or being better than the other. You begin to see our unity and notice how we're very similar.

One mission of Interfaith Power and Light is to work in solidarity with vulnerable and marginalized communities. We start with listening to those who have experienced marginalization. So instead of asking a straight person about what it's like to be a queer gay guy, we learn directly from members of that community. The same is true when it comes to social justice issues. When I first learned about the importance of equality between men and women, it rocked my world because I had grown up Christian in a conservative space. Simply put, a man doesn't know a woman's experience and shouldn't be telling her or others how to think about her life or her body. It's a big deal. We all are experiencing marginalization of some sort, and we should listen to each other.

What we find in the environmental movement is that black, brown, indigenous, working-class, and poor people are the ones impacted first and most intensely. The Palisades fire destroyed neighborhoods with multi-million dollar homes. In contrast, the Altadena fire burned down an all-black neighborhood, and so now you have a homelessness issue in Altadena. All the black folks are experiencing this. Evictions are rampant. People can't pay rent right now because of all of the craziness.

Black and brown people are impacted the most by pollution as well. Our office is in West Oakland, which is a predominantly black neighborhood near the port, is so polluted that 94% of the kids who live there have asthma, which means they can't go out and play like children used to be able to. That also means they might gain weight, which in turn may lead to knee problems later in life. So these kids may have health issues for the rest of their lives, all because they grew up near a port where tons of trucks pump pollution into the air each day.

Once I understand that this neighborhood in West Oakland is most impacted, I need to tell myself, “Okay, I don't live in West Oakland and I'm not black. So I want to listen first and get to know the neighborhood and develop trust. I need to earn people’s respect by showing that we are not just going to dip in and dip out.” Then we build that relationship of solidarity, trying to bring ecologically good things to this community that otherwise wouldn't have access to them, whether that's EV charging stations or EV cars or battery backup or solar power. It's things like this that they don't have access to because the structural system has excluded poor people and people of color.

The best way to heal these harms is not to go in thinking you've got it figured out, but to nurture a relationship of solidarity. Show respect, listen, gain trust, and then move forward. At that point, it's no longer “I'm a nonprofit rep here to help you.” It's “We're in this together. We just happen to have the structure of the nonprofit so that we can get grants to make change happen. But ultimately, we are not here to save you. I mean, I need saving too.” I learned this through those relationships of solidarity.

Another one of our values is embracing hope. We live in a situation where the future does not look rosy, whether it’s climate or the current political environment. But religious traditions speak very strongly about hope. We have morals and ethical standards. A favorite quote of mine is “The only thing stronger than my fear is my conviction.” So I am terrified of the future, but I know to my deepest core that I need to act. I need to take action out of respect for the interconnectedness of all beings, out of respect for my neighbor. In doing so, I bring hope into those spaces. It is easy to bring despair everywhere you go, but what good does that do? People aren't motivated to act out of despair. People act when they're inspired by great feats of human action for a better world.

One of my favorite examples of a real-life situation was a protest at a Wells Fargo bank. I participated in many of those, shaking my fist and yelling and screaming about divestment. Divest, divest from fossil fuels! Here's a flyer, divest! And those looking on are thinking, "Oh my God, who are these people?" And security comes over, wanting to arrest you or kicking you out. And then it gets aggressive, and I end up asking myself, "What in the world happened here?"

But I've also gone in with a bunch of pastors urging divestment from fossil fuels. We sang a fun song. And then everybody all of a sudden is looking and asking, "Who are these people? What's going on? This is cool." And now they're taking flyers rather than being handed flyers. Now security comes over and they're curious, "What are y'all doing? This is hilarious." So you have shifted the narrative. It's because these people are now filled with hope and want to share that spark of joy. Those who sang were not your typical burned-out activists but were tuned into something, tuned into spirit you might say, that gave them that vibrancy. To be in the middle of a Wells Fargo with a bunch of people filled with hope! The impact we had there was so much stronger and so much more interesting and fun and long-lasting than when we just went in with anger.

There's something about religion that says, be righteously angry, but do it with hope, do it with joy. There are creative alternatives to just yelling and screaming. Already in my early Christian days, I was inspired by Jesus offering this sort of third way of seeking peace, not by ignoring the world's problems but by finding and navigating these weird and interesting paths in between. He often used street theater and other creative ways to bring people in who otherwise wouldn't be convinced. I find inspiration in leaders like this. It is probably true of all the great religious heroes—Muhammad, Buddha, Krishna, Abdul Baha. That third way is not "Shame on you" or "You're bad." It's being merciful and gentle while still speaking truth, like the Bishop of Washington Cathedral when she confronted President Trump.

Let me say some more about the work CIPL is doing with congregations. We come at this from three directions, promoting changes in narrative, policy, and facilities.

One thing is that we try to change the narrative. Storytelling is really important, and religions are full of storytelling. We try to find our personal life stories within these ancient traditions, and we tease out the parts of hope, the parts of ecology, the parts that speak to climate justice.

We help people recognize that it's not their fault that the world is falling apart because of climate change. The real story is that billionaires and corporations in these massive systems are causing this, not the long shower you may have taken this morning. Rather than blame individuals for their acts, we need to change the system—and we can do that together. So we change the story in that way, rooted in our wisdom traditions, because for thousands of years, our religious traditions have been telling us to take care of each other and of creation.

Another approach is through policy. If we want to make some sort of structural changes outside of social movements, we can work in the policy realm, trying to activate communities. How do we get involved in the political process? To be honest, before I joined CIPL, I was very engaged in activism, but I wasn't following legislation the way that I do now. I didn't know the power of interacting with legislators. Now I go into a congregation and tell people that they have quite a bit of power in our system. And so let's leverage it. Let's write those letters. Let's call our senators. Let's vote for the right bills. Let's make sure that they vote for the right things. They're not always going to listen, but we can try our best to reach policymakers.

One of the cool things about congregations is that they meet in buildings. Those are good spaces for collaboration. We can flip those buildings into spaces that become “resilience hubs.” Congregations have this space that can provide solar and battery backup so that when the power goes out, your neighbors can come to your facility and feel safe. I am diabetic, and people like me need to charge up or need their insulin refrigerated. If a church has backup battery power, I can go put my insulin in the fridge. And so the church or synagogue or temple is now swinging wide its doors open to the community, saying, “We want to be a resource for you when climate disaster happens. And so we are preparing our building in a way that's going to help mitigate and enable us to respond.”

There are other side benefits. It’s hard and expensive to get some of these changes implemented, and so you learn in community how to do it together. Then, as a next step, you can take what you learned and apply it at your own home. So it's not just the big facilities that change.

So those are the areas of changes we want to effect: the stories to tell, policies to support, and facilities to adapt.

Many people claim to be spiritual but refuse to have anything to do with organized religion. Therefore we have started to shift our language to people of faith and conscience, to include our atheist friends. I've worked with Atheists United, which is a group down in Los Angeles. Atheists United said, “We don't like the religious stuff. We just want to build community. We know the world's crumbling because of climate change.” Because of that common goal—building community and preserving the environment—we were able to work together.

Personally, I believe that we are called to change the world for the better, not just think about it. I have found it difficult to change the world when you're just off by yourself, running around doing random things. However, if you're part of an organization of some sort, there's historical memory, there are leadership opportunities, there's accountability, there's so much more. As much as I love my spiritual but not religious friends, for me, it's an easy cop-out. I want people to be organized into religious traditions because there's more power. We can do more. Once you have a facility, you can do things. You can hold meetings, you can host people, you can house people, you can feed people. You can translate all of these visions, values and ethics into action.

I wish I could bring people of different faiths together. In my work so far, Eastern religions have been present, but not as present as Christian and Jewish traditions. To me, the world would be more whole if we could find some path toward collaborating more, some way of building bridges between religious traditions. There should be a website that says, "Come and tour our religion. If you're a Christian, here's how to act in our temple. Come see, come explore." We could really expand and broaden our horizons. For example, learning about Hinduism and yogic traditions totally transformed my life. Yoga invites us to sit still for five minutes with our eyes closed and notice how hard that is. Just to be you, just to be, and not to try to think about anything else. There's power and peacefulness that can come from it. For so long I was searching for God "out there" when I could have searched inward and found peace as well. This whole religious tradition developed around being in tune with one's body—I thought that was the coolest thing in the world. If we could engage more of that at a larger scale, I think the whole society would benefit.

For Further Reflection

- ❖ Do you consider yourself a religious person? If so, how does your faith impact your stance on climate advocacy?
- ❖ What features or teachings of other religious traditions do you admire?
- ❖ How do you go about building trust and solidarity, both person-to-person and in community?
- ❖ What did you learn from Greg's story of West Oakland or the story about the Wells Fargo bank protests?
- ❖ Which of Greg's insights spoke to you the most and will likely impact your own life?
- ❖ If you are part of a faith community, how does it interact with the vulnerable and marginalized?

Jan Warren

I've been involved with California Interfaith Power and Light for probably ten years now. It began when the organization shared these little fluorescent light bulbs with congregations. I later became reconnected when I joined a newly formed faith climate group.

My activism, whether it's related to housing, immigration, anti-racism, or climate, all springs from my faith. I've lived my life through the lens of Christianity, based on the teaching of Jesus that when people are hungry, naked, sick, imprisoned, or strangers, we respond to them and help them. I'm focused on who is being harmed. Who benefits from whatever the action is?

Who is at the table, and who hasn't been invited to share? Who makes the decisions?

I don't remember where I first heard the expression "Pray with your feet." We are not called to be perfect, but we are expected to show up. I feel very strongly about doing my part. With all the craziness in our country today, I need to stay grounded in faith. The inspiration for climate work is the interconnectedness and sacredness of all life.

I have lived my whole life as part of a faith community. There are many false prophets, and my pastor has taught me to discern what's true by using the arc of scripture alongside our lived experience, tradition, logic, science, and intellect. So you can't just pull a verse out of scripture to make a point. You can't simply quote a verse about a murderous warring king and use it to justify war or killing. You have to follow the whole trajectory of the Judeo-Christian tradition. Jesus was born, lived, and died in Palestine. He was a Palestinian Jew under the Roman Empire. And so the arc of scripture leads us to ask, "Where are we going? What are we reaching for? How can we do better?"

I am part of the Lafayette Christian Church (Disciples of Christ) congregation in Northern California. After we found out about Interfaith Power and Light, we discovered our denomination had a program to help green the church. We started making improvements around the property. We reduced our energy consumption and got serious about recycling, composting, reusing, repairing, and reducing waste. Over time, we joined MCE (Marin Clean Energy) and added solar and heat pumps. Most people engage in climate action because at some point they recognize that something they love is changing, disappearing, or being harmed. We are called to take care of all creation. We have been given a bountiful gift, but often we misuse it. The connection to faith is that living things are being harmed, and so we have to speak up.

Some years ago we showed Al Gore's film, *An Inconvenient Truth*, at our church. Shortly after that, the Pope's encyclical was published. We held a meeting with Father Bonacci, who is a peace activist here in Contra Costa County, and had a great discussion about the Pope's encyclical on the environment.

The first march I participated in took place in Oakland. I don't remember what it was about. I didn't have a sign, but as is typical for rallies, we had a speaker and we marched in the street. I went by myself and met some people. At marches, you always run into somebody you know and you always meet somebody new. I gave a $2 donation in support of a nonprofit called "Sunflower Alliance" as a sign of support for a group that had formed after the 2012 fire and explosion at the Richmond refinery.

At the rally, I became really curious. I asked Sunflower Alliance members if I could join them, and they graciously invited me to take part in their meetings. At the time, because of the refinery explosion and fire, they were focused on moving the Air District to have better enforcement, better oversight, and better regulations to protect the health of the residents. Each of California's 35 local Air Districts is responsible for monitoring and improving the air quality in their district. And so I began attending the Air District board meetings responsible for the nine counties in the Bay Area. I learned about the harm from volatile organic compounds and fine particulate matter.

The Sunflower Alliance is still active in the City of Richmond. It is a leader on climate issues in Contra Costa County and partners with coalitions around the state. I had to learn a lot, and I did because I had never spoken in front of a regulatory body before. When something came up for a specific vote, we spoke up using facts and stories by others with lived experiences. Improvements were slow because the fossil fuel industry fights everything. So it takes time to create change. Once rules are amended and approved by the directors of the Air District, the refineries sue. Then the process is in the hands of the lawyers.

We continued to grow and learn and have successes. One opportunity was to participate in a "toxic tour" by bus to visit areas of Rodeo and Richmond near refineries. We have a big environmental justice problem because all these plants are in locations where they harm the health of people who live nearby. The fossil fuel plants were deliberately built in unincorporated areas.

I remember going out organizing, knocking on doors in the Rodeo community to tell people about an upcoming event. At the time we were addressing the environmental harm of tar sands: the thick, dirty, awful, bottom-of-the-barrel fuel coming in from Canada. The plan was to ship the tar sands to be refined at the Phillips 66 refinery next to the Rodeo community.

One woman I knew had lived in Rodeo for ages and is up in her 90s now. She said, "I can walk down this street and tell you who's died from what kind of cancer and how it got worse after a different type of refining was used with even more harmful chemicals."

So we knocked on doors to encourage people to come to the meeting and to let them know we needed help to fight against the tar sands. Everyone who lived in the community knew the smell, black smoke, and flaring that came from the refinery. They were aware that many students needed to bring inhalers to school as a consequence of pollution.

The members of that community sometimes get discouraged because the fossil fuel industry is so powerful. Big Oil acts like they can do anything. By donating money for the baseball team, local schools, and other areas of community need, they then argue, "Look at how we're helping you." So the people find themselves between a rock and a hard place because they want to count on that kind of "help." They know that they are living in an area that makes people sick. Refineries also fight every regulation from the Air District that would improve their health; it's cheaper to support local organizations and pay fines for violations than to invest in the health of the community and thereby reduce their profits. The good news is the tar sands project eventually was stopped anyway.

There is a lot of "greenwashing." One Coke commercial shows their soda pop in a glass bottle. The reality is that Coke uses glass bottles in Mexico and reuses them; however, in the U.S they mostly use plastic bottles.

We are bombarded every day with advertising, selling, selling, selling. It is up to us as consumers to reduce the waste of food and clothes that we overbuy. The textile industry makes an excessive amount of clothing that at the end of the cycle through markdowns still ends up in landfills. Clothing made of polyester is another product of fossil fuels. By reducing our consumption and embracing a circular economy, we can contribute to a healthier planet for those who come after us. We can embrace materials that can be recycled, like paper, cardboard, and aluminum.

In the 40s and 50s the fossil fuel companies were looking for a new product to grow the market. The bonanza turned out to be single-use plastic products sold as convenient and disposable. The trouble for the environment is that making more plastic items tends to be cheaper than repurposing existing ones. In addition, all plastics leach chemicals, especially when heated or scratched. Plastics don't go away, they just break down into smaller and smaller "microplastics." We have grown dependent on these products and packaging because they are convenient or cheap, without considering where they end up once we no longer need or want them.

Fortunately, since 2015 more and more climate organizations have formed and thousands of people have gotten involved in the environmental movement. You can't live in California with fires and heat, droughts, and floods, and not be aware that things are out of whack. We all can make different choices like driving or flying less. There is no end to the ways we can help, it just takes starting a new habit and loving the planet we live on.

This year on Presidents Day, I saw the largest rally at the state Capitol. It was so cool. We found this little non-corporate coffee shop where we stood in line to use their one bathroom. Some of the rally organizers furnished markers and poster boards, inviting people to sit and make posters. By the time we left the coffee shop, the crowd was huge. People were everywhere with their signs. There was no trash. It was peaceful, and you could feel the positive human energy. People who drove by in cars honked horns to show support. I was filled with hope that day. We have made progress in many areas: filtration systems for people's homes, replacement of gas stoves, pollution reduction through clean energy alternatives.

Fifty years ago in eastern County Contra Costa, they were drilling oil wells. Then they stopped doing that because they could find more productive places elsewhere. People started building homes nearby. Some of them had a vent for the methane that was coming from underneath their home. People who bought these properties had no idea that their house was on an old oil field or oil well that was leaking. Sunflower Allowance is now suing the California Geological Energy Management Commission. People want to have a say in whether anyone is allowed to drill in their neighborhoods. It shouldn't be permitted just because someone wants more gas or oil. We should be in control of our health.

The more we work together, the more we can accomplish and the more people can have healthier lives. Now physicians, nurses, scientists, artists, musicians, and all sorts of folks show up to help. I think we learned during COVID how important and how life-giving it is to be outdoors. Nature keeps our minds and spirits uplifted.

I want everyone to be in an environment where it is healthy and safe to live.

The scientist Katharine Hayhoe from Texas Tech University taught us how to talk with people about climate. She advises us first to find a common connection between you and another person. Everybody loves something. Find out what they care about, whether it's their grandkids, their health, or being outside to play golf or pickleball. This isn't a quiz. It's an opportunity to really get to know someone and find something you both love. Then, make a connection to climate. That's not hard to do. Some folks you're never going to convince until they experience a climate disaster themselves, but I believe most people want to be helpful; they just aren't fully aware of the problems or have not thought about climate before. Look at the mutual aid that pours out when there's a disaster. The government can't fix all that. It requires all of us to pitch in. That's how I see my role too. It's why I keep doing what I'm doing.

For Further Reflection

- How does Jan link her activism to her faith? She refers to Matthew 25. Is that passage a guide for you as well?
- For her social justice initiatives, she focuses on people who are harmed. She sees that focus as part of the overall trajectory of the Judeo-Christian tradition. Do you agree?
- In 2015, Pope Francis published his encyclical on the environment. It is available online. Read and discuss it with a group of friends.
- Have you ever gone door to door to inform or persuade people? How did you talk to them? What did you advocate for, and what did you learn?

Will Morris

My eco story started early because, without realizing the implications, I grew up in a town right down the street from a refinery. I lived there, not knowing any better. Apparently my neighbors were used to it and thought it was normal. Often in school we had to run inside and shelter in place because gas would escape from the refinery. That happened all the way from elementary school through high school.

Another part of my eco story is that during my childhood my parents took me to national parks in California, including Yosemite, Sequoia, Death Valley, and Joshua Tree. I enjoyed the outdoors in all these beautiful natural places. At home we would watch things on PBS, like *Nature* and *Nova*, and that created an appreciation of science.

I grew up in a Christian church context where I was taught to love God and to love my neighbor. In my young mind, that love also extended to caring for our Earth and making sure that our neighbors were taking care of the environment as well.

It wasn't until high school that I got clued in on climate change specifically. I was lucky to get to take an AP environmental science class with a super enthusiastic teacher. He taught us about biomes and ecosystems and all these wonderful things on our planet, and something in my brain and my heart just clicked. I knew that I wanted to spend my life studying, enjoying, and supporting the environment.

At the same time, my teacher dove into the science of climate change and what was happening. Our high school was only one block from the refinery. He pointed to the back window of our classroom where we could see the refinery, and he talked about how the burning of fossil fuels contributes to pollution and how we are suffering from the effects of refining oil because we live right next to where it's done. I found it very interesting and wanted to learn more about it.

When the time came to decide where to go to college, I chose a place that was environmentally much more attractive than my urban, concrete-covered home town. I went to Humboldt State in Northern California, in the middle of the redwood forest, and majored in environmental science. Humboldt State was a beautiful place to be in nature, learn about God's creation, and study how to take care of the Earth.

At the same time, I was wrestling with my faith. The church I grew up in had taught me to love God and my neighbor, but it was also conservative evangelical. The members thought that environmentalism was not for them because they mistrusted science. It was on the other side of their political spectrum. So I felt torn between identifying as Christian and pursuing what I wanted to do. Fortunately my parents kept encouraging me to explore environmental science. Their positive reinforcement enabled me to continue my studies and feel supported. As time went along, they too began to feel a bit conflicted and started surrounding themselves with people who had an open mind, who showed love of neighbors and community rather than focus on individual shame and sin. They joined me in this journey at their own pace. I now attend a United Methodist church where people want to talk about ecojustice and creation care.

At Humboldt State, I was one of very few Christians majoring in science. I wondered why so many people thought that faith and science were irreconcilable even though they felt holistic and integrated to me. I felt lonely at times. Those who had a strong faith didn't care about the science part, and vice versa. And I found myself in the middle.

Many people think that the teachings of the Bible contradict science, not realizing that the Bible was never intended to be a scientific book. It is a story of people on a faith journey. But the church I grew up in was fundamentalist, insisting that each word in the Bible had to be taken literally. They select different verses in the Bible to prove their point. So when the Bible talks about "dominion," that justifies what we do to the planet. The science is clear about climate change, but people don't want to hear about that. They want to hear what the Bible has to say about it.

Interestingly, the Bible can be the biggest tool to convince people to be environmental advocates. When you zoom out and look at the context of those verses, you find that God is calling everything good and wants everything on Earth to flourish. God is revealed through creation, and we're here to take care of it and to steward it. Our original purpose, in the Garden of Eden, was to be in a healthy relationship with the Earth and all the creatures on it.

The church I grew up in never talked about environmental stewardship. So now, as climate advocates, we can use the larger context expressed in the Bible and show that it's all in there, some of it even in the first few pages. It occurred to me that I was supposed to help bridge the gap between environment and science on the one hand and faith on the other. I had learned about both sides independently, and now I was able to function as a kind of translator, using the language of both. I can talk about the policies of climate change and at the same time mention creation care, stewardship, and taking care of our Earth through the systems God gave us. I function as a messenger, a go-between. Gradually, people in church who were kind of skeptical are becoming more receptive. They wonder why they hadn't heard this message earlier. Now they open their Bibles and see that it's right there. They begin to say, "Oh, you know what?

You're right. Why haven't we been thinking about this more? We should integrate creation care into our lives."

After I finished school, I tried to figure out what to do next with my life. The church I was attending at the time had a missions conference where I heard a speaker talking about doing environmental work as a Christian. That was the first time I'd ever heard of that option. I told him, "I just graduated, and I really want to do this." I ended up joining some of these mission groups, traveling to Kenya, Chad, Indonesia, and Mexico to do environmental work.

Climate change and environmental degradation have affected these places in a number of ways. Members of our group wrote reports to relate how communities were affected and proposed solutions and funding for specific projects.

For example, in the country of Chad, which is in North Central Africa just south of the Sahara Desert, the soil is very dry and rocky. We learned from local people, leaders, and village elders that the rainy season had shifted by an entire month. 98% are subsistence farmers who know the rhythms of the land that their ancestors have lived on for thousands and thousands of years. Suddenly that land has changed and it's hard to predict and less stable. Crops were planted when they were supposed to be, but the shift in the rainy season caused floods. Conversely, when rain was supposed to come, they found themselves in a drought instead. There was also instability in the amount of rainfall. Instead of being regular, there were flash floods, and so the water didn't go down into the ground and the wells ran dry.

Many of these things, like the shifting rain patterns, weren't fixable, so we collaborated with local folk to discuss what adaptation and resilience might look like. Would there be ways to store more water? What changes could people make in water usage and farming? Some international folks worked with people on the ground who had farming experience and were familiar with the villages in the surrounding areas. We listened to their stories and turned them into data. We asked, "What has changed in your lifetime? How did things use to be and what have you seen?" The locals were in tune with the land and often knew what was happening way better than I do here in the United States where I don't have to grow my food. They told us how their parents and grandparents had lived and how much had changed in just the last couple of decades.

We wrote up our reports for the mission organizations who in turn sent people with training and skills to help meet the needs overseas. It didn't matter whether they were Christian or of other faiths. In their minds, everything was whole, like climate and environment and their faith; all were together as one. In these different places in the world, people knew that climate change was happening. They knew they were supposed to care for the Earth. It was an integral part of their faith.

When I came home, I talked in different churches about what I had seen and learned. Some would say, "Wait a minute, you're Christian and do environmental stuff?" It was still the same questioning I had heard in my early church community. I didn't encounter it in other places around the world. The people there got it. They thought in a more holistic way about how to live out their faith by applying it to their everyday lives.

I had moved back to Southern California after graduating. I wanted to talk about the environment and creation care and climate, going as a volunteer to as many churches as I could and talking with the pastors or young adults. At that time, most of the congregations didn't have creation care team programs or messaging or sermons. I wanted to change that, to convince them that the environment was important and that there was a theological basis for acting to preserve it.

At one point I found out about an organization called "Young Evangelicals for Climate Action," a nonprofit here in the US that promoted the exact thing I was doing. They educate young people in congregations about climate and about how to reach church members. Once I found them, it was a lot less lonely. It felt like I had a group of people with similar goals, working in different parts of the country. That was wonderful. They educated college students as well as youth earlier in their journey.

Young evangelicals deal with the conflict between their allegiance to planet Earth and what they're currently being told in evangelical circles that support right-leaning politicians. They see the contradictions. In the past 10 years, most young people have gotten their information and formed connections on social media. Even those in conservative or evangelical spaces believe in climate change. For them, it's a no-brainer. They are trying to bring their insights into the churches they attend, saying, "This is happening. We can see it with our own eyes and we should be adjusting to it." They might get responses like "Those ideas are from the other side" or "You're being brainwashed." But a shift in consciousness is going on. Most young people have opened their eyes and are recognizing the climate problem. A few have doubled down and said, "This is not something we want to get involved in. Those are not our values." Like with the adults, the divide has gotten even wider over time. Some of these young people who are in conservative evangelical churches are really stuck.

My role is to become a trusted messenger for those places. A person who comes in as an outsider has a hard time gaining people's trust, but if it's a young person who grew up in that church and people held them in the nursery as a baby and their families stayed involved, then that's a different matter.

In the climate space itself, groups of young people are coming together and forming what seems like a church. They talk about spiritual things, like how our soul is doing and how we're living out our values. For many of them, caring for the Earth is a spiritual thing. Some may have been religious earlier but were burned by the rigidity of their congregations and don't want to go back to that setting. So they create a faith community of their own. Environmental action could be a wonderful opportunity for churches and other houses of worship to welcome young people back and brainstorm with them about how we can serve our community through climate and environmental initiatives, programs, and goals.

I left the mission organizations because I wanted to focus on the US. I also learned about theologies of justice. Christian mission systems have caused a lot of harm around the world. Growing up evangelical, I didn't know that. But once I learned about it, I chose not to be part of that anymore. Instead, I joined the Young Evangelicals for Climate Action. Now I'm chair of their steering committee, so I've stayed involved with them.

I trained with Al Gore's Climate Reality Project and gave presentations on climate for a while. I also met someone from Interfaith Power and Light in Southern California. They were again doing the same things I was doing, but in an interfaith context. Now I'm on the board for California Interfaith Power and Light and help support their initiatives.

For the last three years now, I've worked full-time at an organization called "Green Faith" which is also a multi-faith climate justice nonprofit, active in multiple countries around the world. I help churches in the US, educating and mobilizing them toward action by linking them to various campaigns in the state. In Southern California, many neighborhoods have to cope with oil drilling and refineries, so we explore what our churches can do to help people who are affected. It has been wonderful for me to work with Green Faith and with Interfaith Power and Light.

I appreciate that it's multi-faith work. I grew up with the most narrow theological worldview, but my conservative and evangelical persuasions have been replaced by openness to people of many different faiths.

It feels like a beautiful tapestry or mosaic, reflecting that we are called to take care of the Earth and that there's spirit in all these different ways of going about it. That openness makes sense to me. I feel blessed to live and work for faith-based climate and environmental justice. Our calling, written in different sacred texts and stories, is to stay in relationship with this Earth and its Creator. We have to do our best to see that all life can flourish.

For Further Reflection

- Did you grow up in a faith community? If so, what were its basic beliefs? Do you still share them? Explain why or why not.
- Do you see science and faith as irreconcilably conflicting?
- Will's religious background was conservative evangelical, but he moved on to a more inclusive faith that is open to other views and religions. Where are you on that spectrum?
- Are you moved to preserve creation because of your faith?
- What makes Will a uniquely qualified bridge-builder for the environmental movement?

Eileen Altmann

Our church, the First Congregational Church of Palo Alto, has several members who are deeply engaged in one way or another with climate action. We occasionally organize events and have had committees working on the topic, although those efforts wax and wane depending on whether a leader is willing to take on the organizational task. But many individuals are doing incredible things, either in their professional lives or in their volunteer efforts around climate action and climate justice. We are aware of each other, pay attention, and feel encouraged.

For example, last weekend we held a women's retreat with the theme of "Sacred Waters—Renewing, Inspiring." It was very climate-action focused, but specifically on water. I look for signs of God's light in the world, and I find it in many places. At the retreat, I learned about a little congregation in Pacific Grove that has taken on a ministry of caring for Monterey Bay. Young people from all over the country come for a "Blue Theology" focused on ocean ecology and the threats of climate change to ocean health. The team offers spiritual practices around this theme and offers hands-on science work.

It was inspiring to hear how one local congregation has taken on this special preservation initiative because of where they are located and because they believe in promoting a healthy environment. This small group of people says, "We can't do everything, but we can be good stewards of this place and share our stewardship with others around the country." The young people who come to help can then take that experience into their adult life. It will be part of their consciousness. They have learned about the threats to ocean ecology from climate change. Moreover, they have learned that there are things that they can do individually and collectively to make a difference.

Another example is a congregation member who helps folks get access to clean drinking water. She was a Peace Corps volunteer and is working with an organization founded by other Peace Corps veterans. So these are folks who ask themselves, "What can we do to make a healthy life more possible for more people? What is an action that brings hope to others?" In her professional life, she encourages and facilitates solar adoption, but as a volunteer, she helps low-income folks in remote areas. The group she gathered at the retreat discussed actions people can take to facilitate access to clean water.

So, even if we're not always well organized congregationally, we are eager to learn from others. There is much energy within our congregation around environmental themes and actions.

Preservation of the environment is central to my faith. We are called to be caregivers. For me, the charge to be stewards of creation does not imply "dominion" in the sense that we get to do whatever we want with the set of natural resources into which we're born. We are very much an integral part of creation and are dependent on and related to all life on Earth. We are just one part of this big ecosystem, and we have a role to play in caring for this gift from God. To me, it's a calling. We need to be aware that our actions are imperiling the health of this ecosystem. As a pastor, I need to heed the prophetic call to change the way we live and to support and encourage the congregation to find ways to work toward a healthier, more sustainable planet.

My concern for planetary health started when I was in seminary. I took an eco-theology class and wrote a paper on water because California has an interesting history around water rights and water usage. Even earlier in my seminary journey, I took a class called "Third World Feminist Theologies." We read about folks who live in less developed parts of the world where life is much more precarious. Reading the biblical text alongside these challenging conditions of life changes one's theological perspective. Third-world feminist theologians speak to the relationship of humanity to the environment. And so I started just paying attention to those kinds of voices.

When I went to El Salvador early in my seminary studies, I could see the threats to the environment there and even heard somebody speak to the ways that climate change was affecting daily life. So the voices of these people, who were definitely more vulnerable than we, started to infuse how I thought about what it means to be in ministry. It translated not only into climate advocacy but into taking a stand on other justice issues as well.

Often when I lead the congregation in prayer on Sundays, I try to express gratitude for safe water, clean air, and other basic necessities of life that I think we tend to take for granted. It's because I am so aware that many people do not have access to safe drinking water or clean air to breathe. And when I pray for folks in natural disaster zones, I lift up from time to time that many of these natural disasters happen because of the climate crisis.

Many people feel climate grief at times, especially when the political leaders who hold so much power are doing everything possible to thwart the kinds of change and activism that so many of us have been working so hard to do. I share their grief, and I understand the feeling of despair. But we have to remember that when there's life, there's hope. If we don't do what we can, where we can, however we can, then change definitely is impossible.

I try to look at examples in history where it seemed like all hope was lost, and then there were people working behind the scenes, unnoticed, unheralded, who were building the foundation for the needed transformative change. I'm thinking a lot these days about all the activism in the African-American communities during the post-reconstruction era and what we call the civil rights era. All kinds of stuff was happening: fascinating writing, fascinating organizing. Even in the midst of devastating racial terror, there were folks who kept insisting, "Change is coming. Change is possible." So many communities were really suffering in a variety of ways, but they persisted, knowing that the relationships they were building and the books they were writing would inspire others and help them tell the stories that needed to be told. There are many such examples where folks were building for a new and better day.

I don't see myself as a charismatic speaker. That's not necessarily where my gifts lie. In my role as pastor, I try to encourage, support, and lift up those who are doing the good work. Networking is something I am good at—paying attention to the various things happening locally, regionally in the state, nationally, and then internationally, and helping build bridges when the time is right. It's not necessary to flood the zone with information, but when there's an opportunity, I like to connect people by saying, "Well, so-and-so is doing this and I think you would love working with them on this."

I do feel like my prayers raise consciousness about injustices and other situations in need of healing and transformation. They also offer encouragement for people who are already prone to action to continue to do the work they feel called to do.

I'm a newly minted grandmother as of six weeks ago. At this point, my advocacy is rooted in commitment to the generations to come. Now that I'm in my 60s, I have a greater awareness of my limited time on the planet. In whatever amount of time I have left, whether it's a day or a decade or three decades, I will try to leave a legacy for the generations that will follow us.

I want people to have a foundation of faith, to know the power and goodness of the Christian tradition. Christianity has always been a subversive movement, because at its essence, it lifts up the last, the lost and the least, the vulnerable among us. So it can be the faith foundation for a better world.

I realize that some of those who call themselves "Christian Nationalists" would be surprised to hear about the subversive aspects of our faith. I grew up in the very conservative Southern Baptist religious tradition. Many of them, but not all, are enthused by Christian Nationalism. But the funny thing is, they do read the Bible, and they taught me to love it, even though the Bible is the most subversive book there is. If you really read it, it's so subversive. Just think about what Jesus said and what he did and who he cared about! Look at how Jesus was interacting with the Roman officials and even the authorities within his own religious tradition! He was very much challenging the assumption that the most powerful will always stay in power.

I know that a lot of young people have turned away from Christianity because they see it as aligned with a toxic power dynamic that they don't want to have anything to do with. That is a crisis for the church. We are losing a generation of young people who do not see the power of the story of Jesus. But at the same time, because there's this subversive element within our tradition, there are young people who are catching on to that too.

We are in the middle of such a mess right now. It's really hard to know what will happen. For me, the cause that most needs a prophetic voice is one that speaks up against white supremacy. That cause is dearest to my heart because it is part and parcel of all the other injustices. We took the land from the native people who had an approach to environmental stewardship that is beautiful. When our European ancestors came here, they refused to learn from them. Their destructive actions stem from a conviction that we know better than these darker-skinned people about how things should be. For the same reason, we abused the labor of people of African descent, insisting that we had a right to enslave human beings.

To this day, some think we can take whatever we want. If we want to own Greenland, so be it. Interestingly, the indigenous people of Greenland have been warning about climate change longer than almost anyone else. They are seeing the effects of global heating more clearly than elsewhere in the world, and they've been trying to sound the warning for a long, long time. Our inability to hear from them is another part of white supremacy. I think because I grew up in the American South, I am particularly aware of that ideology. I understand how people set themselves up to be superior because we are white.

Here is my message for the younger generation: Your creativity, your energy, and your power to envision a better world can be the force of transformation. Stay in conversation with the ancient tradition of your faith, whatever that faith might be, and mine it for its wisdom on how to make positive change possible. Use all the youthful energy and creativity you have, but also draw on your faith tradition. Bring those two things together in conversation to help us discover what we all need to work on together.

For Further Reflection

- ❖ If you are part of a faith community, why is it important to you?
- ❖ If you belong to a congregation, what do you know about their initiatives to address the climate crisis?
- ❖ Have you heard or read about Third World Feminist Theologians? What did they advocate for?

- ❖ Talk about a local effort that for you is a beacon of light.
- ❖ Think of people in history who brought transformative change. How did they do it?
- ❖ Did it surprise you that Eileen called the Bible "the most subversive book there is"?
- ❖ Eileen strongly criticizes white supremacy. Do you agree with her? Where do we find examples of racism today that need to be exposed and stopped?

David Howell

The Anderson Collection at Stanford has a painting by Adolf Gottlieb, a Jewish artist from New York who painted during World War II and later in the twentieth century. He used abstract expressionist forms to convey emotion. "Transfiguration 3" shows two powerful shapes: below, a black blob portraying the chaos and darkness of the world; and rising above it, a bright red powerful light in the shape of the sun, transforming the darkness below. How are these two shapes in conversation with each other? Is one of them winning?

For me, "Transfiguration 3" is a hopeful image. The light floating above the darkness is not a shiny, happy light; it's more a dark blood red. But every time I see it, I think of the phrase from the Gospel of John that there is a light that shines in the darkness, and the darkness will not overcome it. The darkness that we will face in the future with climate change and environmental disasters is not going to be overcome by a gentle light. It will need strength, heart, hope, and collaboration. We have to bring initiative, creativity, and vision to heal the world we love.

The Biblical story of the Transfiguration is about those moments in life when we realize something profound: moments of insight when light breaks into our lives, when we feel a sense of connection through history, and when we feel like we understand a greater truth that can't be defeated or dismissed by others.

Suddenly, we have a sense of who we are and how we're connected to all of life. That's what this story is about, those moments when the light shines and we feel like we can rise above the troubles of the day. Peter, in the story of the Transfiguration, wants to live on that mountain of light and clarity. He wants to say, "Oh, great, let's stay here. I'll make tents. We can all stay." However, we can't live in those moments. We have to come back down to the real world and live in it. On the other hand, in those moments we have seen a light that sustains us, giving us strength and hope to get us through.

Having experienced those moments of clarity doesn't mean we have all the answers. Sometimes we can be tricked into thinking we know it all and that everything we have to say is right. That's not what those moments are for. We make mistakes, and we're going to be wrong. But those moments of light give us memories to draw upon when we feel overcome by the chaos of that swirling mass of darkness shown at the bottom of Gottlieb's painting. We have something to fuel us. It helps us remember that we don't have to be subsumed by crisis and fear.

We all have moments, places, and people in our lives that reveal glimpses of powerful truth and clarity. Think of mentors and friends who have inspired you. I know people who feed the unhoused, who join protests for justice, and who participate in ministries to improve people's lives. The world needs people who are committed to making a difference, especially now, and who gather, talk, strategize, and look for that light in the darkness.

During the COVID pandemic, which was such a powerful experience for all of us, so many people were asking, "How can I help? Who needs me? Where can I make a difference?" That's life rising above the darkness. We see the chaos and the swirling pain and difficulties, and we say, "How can I help?" The darkness is undeniably real, and it feels powerful, but light is shining. We have a choice about where we fix our gaze. We can look at the light or we can let darkness hold us stuck and confused and scared. As we engage with the world, it may not always be safe, but we can be focused on the light and know that the darkness will not overcome it.

A Poem About Earth's Renewal

Our Call to Take Care of the Earth

In the Hebrew Scriptures, humans are put in charge of creation as caretakers. We have proven to be unreliable stewards, contributing much to destroy our planetary habitat. This behavior has contributed to the devastation of the land, led to severe droughts and desertification, and caused the suffering of animals and the destruction of species. Currently, no region of the planet is spared the painful consequences of climate change and environmental degradation caused by our unfaithful stewardship.

I went back to the Genesis account and wrote the reflection below, creating haikus followed by free verse, pondering each day of creation. The "beginning" in the opening stanza is the present. What follows is a vision of the future—the things we could do as more responsible caretakers of the Earth.

Creation Revisited

Like the morning sun
Breaking through pollution haze
Out of death we rise

> In the beginning, Earth and its creatures
> were plagued by droughts, storms, fires, floods,
> starvation, injustice, and disease.
> Confused, we search for solutions.
> We noticed a ray of sunlight
> break through the darkness

and saw the beauty of creation
we had been charged to take care of.
And there was evening and there was morning,
the first day.

Watery blessing
Caresses of soft rain
From sky to deep sea

We began to clean up the oceans,
removed mountains of dumped plastic,
passed laws against overfishing,
helped coral reefs renew themselves.
We limited greenhouse gas emissions
that hung over the land
like a suffocating cloak.
Blue skies returned, and we saw that it was good.
And there was evening and there was morning,
the second day.

Growth on fruitful land
Vegetation yielding seed
Hyacinths in bloom

As allies of the Earth, we planted trees,
prohibited deforestation, stopped poisoning
the land with chemicals, reused and recycled.
We danced under redwoods
and prayed on holy mountains.
We drew strength from the land,
harvesting and sharing what we grew.
And we saw that it was good.
And there was evening and there was morning,
the third day.

Day apart from night
Light in the dome of the sky
Days, years, and seasons

When it became clear
that extracting fossil fuels
would bring certain death
to future generations,
We harnessed the heat of the sun,
the strength of the wind,
and the steadiness of waves and rivers.
We knew energy as gift and birthright—
free, safe, and accessible to all.
And we saw that it was good.
And there was evening and there was morning,
the fourth day.

Filling Earth and sky
Swarms of living creatures roam
Colors multiply

We recognized the interconnectedness
of all living things, not just humans.
Animals, like us, deserved
Care and protection from abuse.
We switched to a meatless diet, ceased
to exploit the essence of living beings
for selfish gain, stopped cruel practices,
and admired the splendor of the species
that co-inhabited our common planet.
And there was evening and there was morning,
the fifth day.

Made in God's image
Caretakers of the garden
Both male and female

We stood together in the fight
against hunger, poverty, and injustice.
We recognized the treasure of diversity:
Black, brown and white, all people together;
humans, animals, plants, all of one substance.

We supported leaders who would protect creation.
Replacing hatred with love,
freed from greed and selfishness,
confessing, we made the turn.
And there was evening and there was morning,
the sixth day.

Earth's wellness restored
Blessings received and given
Time to celebrate

We looked at the now unpolluted sky,
admired the brilliant stars,
breathed the clean air,
joined hands in a dance, rejoiced,
and saw that keeping the garden
was indeed very good
and that we were part of a Spirit
that made it so.

Gratitudes

My sincere appreciation goes to all those who have contributed to the making of this book. I am grateful for the time you took to talk with me and to review your statements to make sure they represented your voice correctly. Any remaining errors are entirely mine. It has been such a pleasure to interview you and learn about your contribution to the climate movement. You are an inspiration.

It has been an honor to get to know you through your stories. Whenever I read them, it feels like I am accompanying you on your journey. Thank you for describing your actions and your vision for a safe environment. Together, we have created a mosaic of life and an image of a sustainable future, each contributing our valuable piece.

As I listen again and again to this ensemble of Earth guardians, I carry each voice in my heart for a while, one at a time, allowing it to resonate with my own life. I admire your valor and your passion, asking myself what I can learn from you and what I can apply as I consider my own next climate action. I take a deep breath and meditate for a while. Often, after reading one of the contributions to the book, I go on a walk to let options for sustainability sink in more deeply. I appreciate your work and send a blessing your way.

Thank you for continuing your environmental advocacy, sharing your insights with your friends and with the wider world. We will proceed on this path together to create an environmentally sustainable, spiritually fulfilling, and socially just human presence on this planet.

Blessed are those who love the Earth, for it is a sacred space.

Blessed are those who mourn the destruction of our planet and will not be comforted.

Blessed are scientists who gave warnings and now provide guidance.

Blessed are parents and elders who strive to preserve our Earth for future generations.

Blessed are teachers who educate for climate justice.

Blessed are nonviolent resisters who reject business as usual.

Blessed are those who remain kind and compassionate in times of hatred.

Blessed are those who seek companions on the journey to planetary recovery.

Blessed are those who speak truth and take action accordingly.

You are Children of the Sacred, Friends of the Spirit.

With you on the journey,
Ingrid Rogers

Made in the USA
Middletown, DE
24 March 2025

73175140R00164